1007416231

The Power of the Line

The Power of the Line: Metaphor, Number and Material Culture in European Prehistory

By

Aleksander Dzbyński

Translation from Polish
Michał Szałański
Chris Whyatt

The Power of the Line: Metaphor, Number and Material Culture in European Prehistory
By Aleksander Dzbyński

This book first published 2013

Cambridge Scholars Publishing

12 Back Chapman Street, Newcastle upon Tyne, NE6 2XX, UK

British Library Cataloguing in Publication Data
A catalogue record for this book is available from the British Library

Copyright © 2013 by Aleksander Dzbyński

All rights for this book reserved. No part of this book may be reproduced, stored in a retrieval system, or transmitted, in any form or by any means, electronic, mechanical, photocopying, recording or otherwise, without the prior permission of the copyright owner.

ISBN (10): 1-4438-4921-9, ISBN (13): 978-1-4438-4921-0

Dedicated to Dobrawa

Table of Contents

List of Figures viii

Introduction x

Chapter One
Theoretical Foundations 1
   Piaget and the childhood of mankind.. 1
      Criticism of Piaget's ideas.. 4
   Communication and culture as sources of mathematical thinking........ 6
      Numbers and language.. 8
      Ethnological examples 10
      Measuring or counting?... 14
   Stages of the development of mathematics..................................... 16
      Stage 0. What came before mathematics? 16
      Stage 1. Protoarithmetical 20
      Stage 2. Symbol–based arithmetic 21
      Stage 3. Arithmetic 22
   Body, mind, metaphors 22
      Object collection metaphor 25
      Zero 26
      The measuring stick metaphor 27
   Body, mind, and space 28
      Number and space in brain research... 31
   Conclusions: the role of material culture 32

Chapter Two
Ex Oriente Lux or How Numbers Were Invented................................ 35
   Clay tokens 35
      Use context... 38
      Tokens and counting... 42
      Early metrology.. 44
      How were units of measure created?.. 46
      The beginnings of the system of counting.................................. 48
   What about Europe?.. 52
   Conclusions... 56

Chapter Three
Mathematization of Space 62
   Constructions and structures 63
      Settlements and their organization 63
      Neolithic houses in the north 70
      Symbolism 73
      Houses and megaliths 78
      Enclosures 82
   Measuring for the living and measuring for the dead 85
      An omnipresent yard? 88
      Archeoastronomy–(measuring–don't leave home without it!) 91
   Conclusions 94

Chapter Four
The Mathematization of Human Relations 100
   Early exchange systems 101
      The enigma of the first axes 106
      Axes in the Eneolithic period 109
      Macrolithic blades 121
   Conclusions 126

Chapter Five
The Metal Age Transformations and the Introduction of Portions 134
   Mysterious beads of the Cortaillod Culture 134
      Seeberg 135
      Colmar 141
      What were the beads? 145
      Different ways of measuring 146
   Measuring vessels in the Eneolithic? 148
      Beaker societies 149
   Other examples of measuring vessels in the Eneolithic Period? 154
      Measuring beaker–the drinking beaker 155
   The European way–from stick to bar 159
   Conclusions 164

Chapter Six
Summaries: Mathematics and a Material Culture 167
   Between Europe and the Near East 174
      Explanation attempts 180
   Value and accumulation 184
      The Bronze Age–price, measure and the warrior's prestige 189
      The social context of the evolution of the measuring stick metaphor .. 192

Bibliography 199

Index 221

LIST OF FIGURES

Fig. 1.1 Adorant from the Geißenklösterle cave (drawing by Dobrawa Jaracz).
Fig. 2.1 Early Neolithic tokens from the Near East (drawing by Dobrawa Jaracz).
Fig. 2.2 Protoliterate grain measurement system recorded on Mesopotamian tablets (after Justus 1999, reworked by the author).
Fig. 2.3 Protoliterate area measurement system recorded on Mesopotamian tablets (after Justus 1999, reworked by the author).
Fig. 2.4 European Neolithic tokens (after Budja 1999).
Fig. 3.1 A schematic plan of the Poljanica settlement.
Fig. 3.2 A schematic plan of a Linear Band ceramic settlement.
Fig. 3.3 The process of manipulating directions and straight lines took on three basic forms. The essence of these manipulations was building a settlement based on the crossing of two straight lines.
Fig. 3.4 A schematic plan of the long barrow cemetery in Wietrzychowice (Poland).
Fig. 3.5 A schematic plan of the Calden enclosure (according to Raetzel–Fabian 2000, reworked by the author).
Fig. 3.6 Measurements of a Neolithic house in Slatino (according to Nikolov 1991, reworked by the author).
Fig. 4.1 In the Linear Pottery Culture axes were valorised by putting emphasis on their morphology (on the left), while among the later Eneolithic cultures the emphasis had been moved to the tools' length (on the right).
Fig. 4.2 The sizes of Eneolithic axes according to Balcer 1983.
Fig. 4.3 Eneolithic axe–shaped pendants and charms (according to different authors).
Fig. 4.4 Analysis of the dispersion of axe sizes in megalithic graves from Kuyavia (after Dzbyński 2008).
Fig. 4.5 a) The fragmentation model of the blade into four parts, b) sizes of blade fragments found in the graves of the Lublin–Volhynia societies, c) sizes of blade fragments in the Tiszapolgár graves.
Fig. 4.6 The macrolithic blade was both a requisite of prestige as well as a mental image of the metal bar.

Fig. 5.1 Two copper strings found at Seeberg, Burgäschisee–Süd (after Strahm 1994).
Fig. 5.2 Distribution of the weight variable for beads recovered from Seeberg, Burgäschisee–Süd. K2–longer string of 36 beads; K1–shorter string of 18 beads.
Fig. 5.3 Kendall analysis of the copper beads from Seeberg, Burgäschisee–Süd.
Fig. 5.4 The copper beads from Seeberg, Burgäschisee–Süd could be produced from one single rod of copper. The numbers present the amount of beads made from specific fragments of the rod.
Fig. 5.5 The distribution of weight of the Seeberg (top) and Geröllfingen (bottom) beads.
Fig. 5.6 Placement and the analysis of copper beads in Colmar. Certain values (shown in weight categories) have been attached to particular parts of the body.
Fig. 5.7 Examples of beads from Polgár–Csoszhalom.
Fig. 5.8 Metrological analysis of the Corded Ware beakers from Central Germany (after Dzbynski 2004).
Fig. 5.9 Kendall analysis of the Corded Ware beakers from Central Germany.
Fig. 5.10 Results of the experiment based on measuring fistfuls of grain (1 portion: M–men, W–women).
Fig. 5.11 Examples of copper axe fragmentation in Central Europe (1.10, 13: Eneolithic axes of different types, 11.12: Altheim type axes, 14.15: axe–bars).
Fig. 5.12 The distribution of the weight of fragmented Zabitz–type axes.
Fig. 5.13 The distribution of the weight of the copper axes in Central Europe in three chronological periods.

INTRODUCTION

In ancient Greece, the rules of mathematical thinking dominated human language and thoughts to such a degree that they gave birth to the concept of eternal, constant and transcendental mathematical laws, to which we have no complete access. Mathematics became the deepest secret, a source of the purest knowledge and perfection. Pythagoras was the father of this concept, though by Plato it found its ultimate expression. He believed that numbers and geometric blocks were the expressions of a perfect world of ideas which are the cornerstones of the universe, whereas mathematics and perceiving the truth through numbers, the ruler and compass became the basis of philosophy. According to Plato, the relationship between the world of ideas and the real world is similar to that between real objects and their shadows cast on a cave wall. It would only require turning one's head away from what catches everybody's attention in order to see the true nature of things. Mathematic formulae, numbers and rules exist independently of our imperfect beings, which are unable to perceive more than their expressions alone.

When did people learn to count? How did they do it? Did arithmetic develop along with the ability to count and how did it evolve at all? Are our mathematical abilities genetically programmed or are they a product of the culture we live in? Or maybe, as Plato suggested, mathematics exists regardless of us?

In order to answer the questions mentioned above, some clarification should be provided. For example, what is counting? If we define counting as the ability to notice changes in small sets and the ability to differentiate between singularity, duality and trinity, then it is probably an ability we inherited from our animal ancestors. The observation of the behaviour of many animal species confirms that they possess an innate ability to differentiate between small sets and even understand (subconsciously) the basic arithmetic procedures, such as addition and subtraction. These observations are also confirmed by research into brain function among people with language impairment. The fact that they can perceive basic mathematical rules suggests that mathematics exists–to a certain extent– outside language and the ability to assess small sized sets is independent of using symbolic language. This still does not mean that mathematics exists outside our experiences, as Plato saw it. However, for the author of this work it means that if mathematical abilities, similarly to language abilities, are the product of the evolution of man then we must assume that the concept of numbers did not exist in the earliest periods of the development of man and society. There had to be a period in prehistory when humans could not add, subtract or even count, even though they used language to pass on symbolic contents and express the subtleties of this world. Today we may only try to answer the question of when this happened and how long this period lasted, although on the other hand we know that we are not yet able to achieve this.

Being aware of the fact that mathematics is the modern mother of all sciences, without which we cannot imagine how man could exist in a world of societies, it is justified to experience a certain intellectual discomfort when faced with the knowledge that for a very long period of time (in fact for the major part of the history of man) people functioned perfectly well without mathematics and were able to live without the ability to count and measure. At some point in history the basis for mathematical thinking had probably been created and the concepts of numbers had been established, although they still had little in common with modern mathematics. However, the matter becomes more complicated if we assume that the threshold of being able to use mathematical abilities is operating on larger sets, abstracting the concept of numbers as well as the knowledge of arithmetic operations. There is a major mental gap between the ability of concrete counting and the concept of an abstract number, which distinguishes us from animal species. How did it come to this? Research over a long period of time in the field of mankind (anthropology, archaeology, and cognitive science) clearly suggests that the development of a material culture in prehistory was a serious contribution to the mathematization of the human mind. It seems that since man started producing more and more artefacts, developing technology and exchanging objects, the role of mathematical descriptions of reality grew. Hence, from this perspective as an archaeologist in the field of material culture in prehistoric times, I would like to present this story. It is the subject of my book.

In order to tell it I have chosen several examples, which do have their own restrictions. I do not claim the right to present a story universal for the whole human species, although my ambition was also to question that universality, or in other words–to support my views on the idea of that universality. In my opinion its cornerstone is created by the following relationship: body–language–material culture. The examples presented allow us to contemplate a less universal phenomenon. The matter of similarities and differences between Near Eastern and European culture in the period of the development of farming should be mentioned in particular. These areas are the basic sources for the material discussed in this book.

In the first chapter the theoretical issues connected with the human perception of numbers and mathematical relationships are discussed. Moreover, some choices had to be made in it. I discuss these issues from the point of view of psychology, cognitive science, philosophy and linguistics, particularly the hypothesis of the importance of metaphors in perceiving reality. Two metaphors crucial for this work have been discussed with greater attention: the collection metaphor and the measuring stick metaphor. At the end of this chapter I call upon the problem of the influence of material culture on the processes of understanding mathematics. This matter is particularly important for archaeologists who study the development of material culture as well as its relationship with man and society.

In the second chapter I present the history of the development of basic mathematical concepts in the Near East. It is exceptionally well prepared and widely discussed in the world of archaeology and is perfectly suited for illustrating the thesis concerning the leading role of material culture in forming mathematical perception. It is also useful for outlining important cultural differences between the Near Eastern and European areas in the Neolithic period as well as for presenting the question of whether it was in some way a determining factor for societies in those areas. In other words, were clay tokens–an invention originated from Near–eastern societies–also responsible for the development of mathematical abilities in Europe?

The third chapter is focused solely on Europe. In it, our area of interest is the organization of space in Neolithic and Eneolithic settlements as well as other types of structures. I elaborate on the symbolic meanings of all constructions from those periods as well as discuss the universal rules of their design. Many researchers emphasize their symbolic and metaphoric meaning. Settlements, houses and the massive constructions of that time are interpreted as a manifestation and embodiment of the cosmic order, which bears many analogies with ethnological studies. As some archaeologists believe, this "domestication" of space had to bear with it a great amount of symbolic meaning among farming communities. There are two basic elements of these structures: straight lines and circles. However, it is the line which deserves special attention. Linearity requires more thought, from its simple manifestations in the monumental form to its complex use in later megalithic structures. Afterwards, I show that linearity was not only an ephemeral symbol and a metaphor but also a practical tool in building anthropogenic spaces–the linear measure, which relates to properties of the human body. In my opinion all of this together requires a new approach. Only when we see a metaphor in the omnipresent linearity can we understand it properly in combination with the cosmologic aspects of architecture, the role of the human body as well as the concept of number. In connection with this I also discuss the problem of the so called megalithic yard and other derived Neolithic measures while proposing a new, more holistic way of understanding this issue.

The aim of the fourth chapter is to focus on the characteristic features of tool production of the European Neolithic and Eneolithic periods and its role in shaping human relations. I mostly concentrate on axes and flint blades and their relation with the fragmentation process, the *signum temporis* of that era, as well as the phenomenon which is less discussed in literature, namely the enlargement of tools. These observations show that both the fragmentation and macrolithization of tools could be two sides of the same coin, as we are faced here with a manipulation of the length of these socially important tools in different social communication contexts. I try to emphasize their symbolical role in shaping the new concept of value among the early farming societies. In my opinion this role is not without a connection with the contents of chapter three, that is with some concepts and idealizations concerning space assessment in the Neolithic period. Furthermore, if we realize that macrolithic technology was directly connected with the development of early metallurgy then we require a new descriptive language, which will place it in a field of a rationalized communication medium. However, we must be aware that in the initial stages of metallurgy, before an abstract concept of measure had been perfected, macroliths could be the only type of link between two categories of perception: they were physically measurable in the times when the weight of metal was still unknown. Many factors point to the fact that the measuring stick metaphor found a fruitful path of development in the case of macrolithic industries, introducing numerical and mathematical messages between people directly.

In the fifth chapter the further transformations of the measuring stick metaphor are discussed. These are best illustrated by means of small copper objects, e.g. beads, three examples of which are presented. The beads from Cortaillod, Geröllfingen and Colmar present clear metrological structures, which illustrate the complex manipulations of the measuring stick far surpassing the simple understanding of proportion, which was present in the case of macrolithic industries. I shall try to show that the clarity of these structures allows us to provide them in a certain "numerical grammar". Apart from the beads, I also discuss copper axes, which also seem to fall under the process of rationalization of the value of metal. Taking into consideration the latter events in the Bronze Age, the fragmentation of metal axes can be understood as a part of this process. I also discuss ceramic vessels, which complete the presented thesis in an intriguing way.

Concluding these arguments I suggest distinguishing two development paths of mathematization and numerosity in Europe and the Near East–the birthplace of farming. These two categories may be perceived as being opposites: the measuring stick metaphor *versus* the object collection metaphor. As it is known, the wealth finance economy is underlined in the European path which stands in opposition to the Near–eastern staple finance economy–a dichotomy which had been described by archaeologists. In the final chapter I have tried my best to outline the theoretical implications which result from this perspective for European prehistory.

CHAPTER ONE

THEORETICAL FOUNDATIONS

Piaget and the childhood of mankind

It has been noted in the past centuries that in many aspects phylogenesis bears a resemblance to ontogenesis. Following this path, the early stages of cultural development could be treated as humanity's childhood[1]. This idea turned out to have great consequences. However, if this is the case then can observations and studies concerning the development of the growth of children lead us to discovering new facts concerning the prehistory of mankind? This idea has indeed been introduced. On the basis of observing children, psychologists of the 20th century began building models of the development of mental abilities, which were derived from basic perceptions. The birth of mathematics was seen in the gradual process of comprehending logical operations. The discussion which began between psychologists and philosophers, such as Peano, Russel and Frege, also led to the involvement of one of the best known cognitive psychologists of that time – J. Piaget, who used it to explain the psychological basis of mathematics as a derivative of logical operations' development[2].

Let us observe that children use their bodies as calculators in their early stage of learning to count. This method, as we shall see below, is the basic rule of the quasi–mathematical register of events. Most often they stop at their fingers using them to show the amount of specific objects without mentioning the number, which they do not know. However, children are often subjected to the rigors of modern education as well as cultural and social pressure from their very early stages of development. These were also matters which interested Piaget. He was a versatile researcher. In 1918 he completed his PhD degree in biology and then focused on the subject of the minds of little children. His studies revolutionized our way of perceiving the development of human thinking.

---

[1] Morgan 1877; Engels 1884.
[2] Miller 1992, 4.

According to Piaget, children travel a path from egocentrism towards sociocentrism during their personal development[3]. In his studies, both in psychological (interviews) as well as clinical (experiments) methods in interviews he often used the mixed questions strategy, interchangeably asking several standard questions and one surprising question while observing the reactions of his small interlocutors. In their answers he searched for something he called "spontaneous convictions". In the course of these studies he concluded that a process exists which is based on a gradual passing from wider and intuitional reactions to such which were expected on the level of social norms adequate to the given situations. When in the presence of older children, younger children adapted their reactions to the expectations of the older ones, while the behaviour of adults were treated as the highest authority[4].

Piaget believed that everything we know about the world, as well as how we reach this knowledge, is the result of how we act in it, how we interact with objects and how we manipulate them. He observed that children under the age of two perceive things which surround them in a literal "objectified" way. At first they behave in an exceptionally egocentric way and cannot yet see the world from another person's point of view. Only in time do they learn to do this as they live in a society. They also learn to pay attention to objects/things and develop a characteristic addiction to them. They learn to recognize things anew after they lose contact with them for some time.

At this stage children cannot make mathematical references between objects and they do not know that, for example, an apple and pear are two objects. They make quantitative evaluations between terms such as "a lot" and "not many". Direct experience dominates over logical thinking[5]. Only after their 18th month are children able to recognize an object which was absent for some time or has been placed in another context as the same one. In other words, they learn to recognize that some things, no matter what happens to them, are still the same things, thanks to which they can be defined as one of a kind and compared with other ones. Children gradually acquire abilities to organize small objects into collections which allows them to compare and understand all tables, chairs, cars, although sometimes very different from each other, as a single collection of objects similar in their nature. Hence, the ability to group items into collections is acquired culturally through learning. Piaget called this the preoperational stage. Magical and egocentric thinking dominates throughout it, which

---

[3] Gruber & Voneche 1977, 91–117.
[4] Gruber & Voneche 1977, 182.
[5] Miller 1992, 8.

gradually begins to fade in time, while logical thinking does not exist at all[6].

When having the ability to group objects into collections we can proceed to the next stage, which is comparing in order to evaluate which of them is smaller or bigger. These are the first steps on the path to mathematization, even if those are quite unsure. As the mathematician John Barrow reminds us, children are quite easy to fool at this stage. For example, collections which we seemingly enlarge by spreading candies appear larger in the eyes of a child[7]. Piaget also had conducted such research[8]. Determining what is larger and what is smaller is in fact quite a complicated cognitive process. Only when having greater abilities, which is the ability to count, can this problem be dealt with definitely. Only at the age of four–five does a young person achieve the ability to efficiently connect collections of different sizes with numbers, which is the numeral expression of collections. This stage has been defined by Piaget as the operational stage.

At the age of 6–7 further, more complicated abilities are acquired. Two values can be precisely evaluated and compared mentally. It is a new quality which shows that some mental objects appear in a child's mind, which it can manipulate in the same way as it manipulated objects/things. They do not have to be perceived directly (they are seen as through the eyes of the soul). This step opens the path to true mathematics. Children at the age of 9–10 are able to replace objects, and actions performed on them, with symbols as well as using them freely in their minds. Dividing and multiplying becomes possible as purely symbolic actions, detached from their object past. Algebra becomes possible.

At the end of this short introduction it should be said that Piaget was also interested in prehistory. Inspired by Lèvy–Bruhl he formed a concept according to which primitive (prehistoric) societies reached only a pre–operational level. It is said that sometimes he expressed his idea suggesting that the reconstruction of mental stages in the phylogenetical sense, which is the cognitive evolution of man, was his main research target. However, being aware that the reconstruction of this type of development in the case of researching prehistoric societies is very difficult, he devoted himself to ontogenesis[9]. Hence, Piaget's study is a good starting point for this discussion in more than one way.

---

[6] Damerow 1996, 1–37.
[7] Barrow 1999, 260.
[8] Gruber & Voneche 1977, 299.
[9] Damerow 1996.

Criticism of Piaget's ideas

Piaget's development scheme, which is universal for all people, suggests that basic mathematical abilities are based on the experience of a person who manipulates objects. In other words, logical–mathematical thinking cannot be acquired *a priori* from objects in the surrounding world but from the way we manipulate them by the use of the body and how we relate to them by means of our bodies. In time, the process of manipulation underwent such an advanced process of internationalization, even dismissal, which we are not able to comprehend how we are even able to count. Hence, we learned this similarly to speaking, which does not mean, however, that the world has a verbal nature.

Nonetheless, being researchers of the past, we are obliged to assume something more. We must assume that the basic logical–mathematical mental structures, apart from individual development ontogenesis process, also develop in confrontation with the widely understood culture. At this point it is worth emphasizing the culture context, which does not occur for Piaget, or does not find any significant use. However, cognitive abilities, in connection to which mathematical abilities develop, are not independent from the influence of the social surroundings and specific space–time in which they are resolved. Following this thesis we may be tempted to connect the psychological concept with its culture–historical correlate. On the other hand, however, we may still count psychological concepts in our discussion of the past.

In time the researchers divided into two groups: those who acknowledged Piaget's theory and methodology and those who tried to deny it through experimentation. Some of the experiments showed gaps in the Pythagorean model as rudiments of mathematical thinking have been found even among infants. The research showed that infants become accustomed with small collections and perceive any manipulations performed on them from outside, as if they had some sort of mathematical precognition. The widely understood criticism made by later researchers slightly shook the theory of the numerical perception development presented by Piaget, although its assumptions generally remained unchanged[10].

The view that the concept of numbers is in some way dependant on the development of logical thinking, as Piaget suggested, was also questioned by Wittgenstein, who claimed that mathematics does not require any logical basis. It can be defined as a "motley collection of language games".

---

[10] Miller 1992, 12.

Wittgenstein promoted the view that the concept of numbers does not necessarily have one cognitive source and that there is no one source of mathematics[11]. He expressed it in typical fashion for him: "*teach it to us, and then you have laid its foundations.*"[12]

In such a general form this view might be accepted from the perspective of this publication under the condition that this study would last a millennia. As a matter of fact, nothing stands before understanding a philosopher's words. However, Wittgenstein's problem is very useful for outlining one of the main levels of our research. We may say that it is included in evolution itself, which occurred in his mind. The early Wittgenstein, present on the pages of the *Tractatus Logico–Philosophicus*, aimed at proving the direct link between language and reality. The main idea of this thesis connected the meaning of an affirmative sentence with its correctness so as to correlate the sentence with the facts in a sensible way. It seems that at times Wittgenstein saw gaps in his early understanding as in the later *Philosophical Investigations* he replaced this concept of meaning with the idea of the "conditions of affirmability": a sentence is sensible if certain circumstances occur which justify its utterance. And, as Kripke pointed out, one of these conditions is inter alia a place in practice of social communication, in the "stream of life"[13].

In time even *Philosophical Investigations* became the subject of a certain correlation (although, according to some interpreters sometime later[14]). Kripke, who dealt in detail with this problem, formed a conclusion which is interesting for our discussion in two ways. One of them I shall present now, the other in chapter 3. Kripke stated that Wittgenstein's view concerning the way names gain their correlation is incorrect as there is no sense of either speaking of the meaning or following the rule in an individual perspective[15]. The correlation for the names (e.g. number) which employ rules are not defined by some definite identifying marks, some specific properties of the world, to which the form relates. This problem is most graphically illustrated by the mental model of the twin Earth presented by Putnam, who wants to show that even when two people speak of the same thing it does not have to mean the same thing[16]. In such a case, this correlation indeed seems only to describe the fact that the speaker is a member of a society which uses a certain term or rule in a

---

[11] Wittgenstein 1974, 297.
[12] Miller 1992, 13.
[13] Soin 2001, 166.
[14] Ibidem.
[15] Soin 2001, 167.
[16] Putnam 1973, 699–711.

specific way. In other words, the norms and rules are understood not only by the use of cognitive categories in the Cartesian or Kantian style but make sense only on the basis of some external system of meanings, passed on by means of tradition [17]. However, as Maciej Soin points out, Wittgenstein had been misunderstood by Kripke, as inside relations, relations between the rule and its use, should not oppose socialization. The rule is not something separate from use, since to understand a rule is to be able to use it. Wittgenstein was aware that it is not an individual mental experience but behaviour regularity, in time transforming into a norm, which forms the basis of language[18].

From this short overview of the voices of criticism we may thus far conclude that the situation is slightly more complicated than it would logically seem from Piaget's experiments. Perhaps children learn from representative symbolic systems before they are completely familiarized with the concept of what a number is and what are its various uses? Maybe they possess something similar to mathematical precognition? On the other hand, the tools used to represent numbers and the ways of using them in a particular society clearly influence the mathematical abilities of individuals. Moreover, we cannot forget that manipulating objects creates the basis for creating rules for speaking about them, as a result of which, at the same time, creates an initial base for socialization resulting from the criticism of Wittgenstein and Kripke. In other words, a model based on one source and one path of development does not explain much, although its main assumptions could be generally correct.

Communication and culture as sources of mathematical thinking

The transition from psychology to linguistic philosophy initiated by Wittgenstein leads us to the sphere of human communication issues. We cannot allow ourselves to present the enormous increase of knowledge concerning this subject which has occurred since Wittgenstein's time. The achievements made by Jürgen Habermas and Michael Tomasello are in my opinion particularly important for the idea presented here. Let us begin with the latter.

Tomasello's theoretical proposition is a link between evolutionism and the humanities [19]. According to him, language and the mathematical

---

[17] Kripke 2001, 147.
[18] Soin 2001, 174.
[19] Tomasello 1999; 2002.

abilities of man are a product of historical and ontogenetic development, which make use of the currently "available" cognitive abilities of the human mind. Some of these abilities are common for all primates and some only for humans. For example, the abilities we share with all primates are "scenes of joint attention" which could have been the initial potential for the first "verb islands", that is concepts and words describing basic types of human activity, such as moving, picking up, passing objects, etc. Hence, Tomasello states that, from the cognitive perspective, language is more primal than man's mathematical abilities. Because of this it is such a symbolizing display of man's abilities, which originated directly from actions that both require attention and communication actions[20]. These in turn, as Tomasello states, originate from understanding other beings as intentional entities. However, a certain paradox is inhered in it, as language, although it is an exceptional ability, is not as exceptional as it is accredited. Jane Aitchison, who researched the prehistory of language, expresses a similar view[21].

Tomasello sees mathematical cognition as an activity which requires gaining, changing and coordinating perspectives originated from social interaction and discourse in order to develop. In the model situation which he presented, this clash of two perspectives results in finding an appropriate solution to the problem. According to him it is also by no means a coincidence that the concepts of abundance and relation characterize language structure. Creating categories and classes of objects (paradigmology) as well as relating them to each other as elements of a sequence (syntagmatics) is specific for language[22].

Habermas[23] presented a similar view and, despite the fact that he was not dealing with the process of creating abilities of mathematical comprehension of reality in the past, on the basis of the theory of communicative action and Tomasello's research we have grounds to presume that the abilities to count and measure are strongly based in an environment of social cooperation and discourse, an environment of language communication. The tendency to avoid the misunderstandings outlined by Habermas could have played an important role in this process. Such a way of communication indeed seems typical for humans and it should not raise doubts that the ability to determine quantity and number precisely is an important element of avoiding misunderstanding in some situations. The philosopher imagined this process as a development from

---

[20] Tomasello 2002, 277.
[21] Aitchison 2000; 2000a.
[22] Tomasello 2002, 250.
[23] Habermas 1981; 1999.

strongly ritualized mimetical communication (with the use of gestures and mimics) to verbal communication, when the basic characteristic is the tendency to rationally reach an understanding[24].

Habermas found that human communication is most effective when seeking an agreement verbally. Language itself bears the rationality potential specific only to the human species, whereby development and actual progress becomes possible. It is suffice to say that the language model of reaching agreements would also play a significant role in forming mathematical and metrological truths. Their origin would involve mostly communicating through gestures and rituals, while further development was based on more abstract reasoning, such as the one present in modern mathematics. It is difficult to say what the end of this process looks like, as this subject is a different story, although Habermas emphasizes that man remains man as long as he retains a "primal" language model of reaching agreements in culture, along with the emotional experiences of the world[25].

The second factor for developing numerical concepts in human communication was defined by Habermas as steering media. Power, and its complex relations, is the first of them. However, in time other measurable steering media appeared, money being the best example here[26]. It is obvious that before its invention, societies had to struggle to reach mathematical claims which were necessary to use money.

Numbers and language

The concepts of the origins of mathematics presented above clearly suggest that its early stages were much more based in language. Perhaps they rather resembled the grammatical properties of language in their structure than their modern, highly abstract, equivalent. The concepts were being executed by creating specific expressions on the same basis as producing sentences in any language with an appropriately complex grammar. Research concerning children's abilities to perform these tasks seem to acknowledge that learning to count is based greatly on the development of lexical abilities[27]. While discussing the idea of man acquiring numerical systems, which had a set counting base[28], Pollmann

---

[24] Habermas 1999.
[25] Habermas 2003.
[26] Habermas 2002, 474.
[27] Tomasello 2002; Polmann 2003, 1–31.
[28] The base in the decimal system is the number 10.

reached the conclusion that the following elements make up the basis of this process:
1. The ability to form rhythmic utterances which had a certain hierarchic structure based on the rhythmic recursiveness of grammar.
2. The ability of words to mark specific beats.
3. The properties of the base itself, which enable the handling of the specific hierarchy of the rhythmic structure as well as the introduction of specific recursive[29] grammar for new numerical words.
4. The ability to reduce the hierarchization of the rhythmic structure.

Moreover, Pollmann pointed out that the process of learning to count/measure is also an extension of vocabulary and grammar. If we were to learn numbers by heart we would never learn to count as the sequence of symbols is infinite. During the process of learning to count we are faced with the problem of generating infinite numbers of utterances by means of a finite vocabulary. In other words, counting is identical to discovering a specific numerical grammar–a lexical structure which enables us to effectively manipulate symbolic expressions. We are not only able to produce the numerical expressions of an infinite length but also a certain order which defines the meaning. This order results from possessing a counting base. The sequence of concepts of numbers which our numerical grammar can produce are nothing more than lexical units. Hence, numerical grammar is also morphological grammar[30].

The possibilities presented in points 1 and 4 are biologically possessed by every human being. However, it is beyond doubt that point 3 mentions a certain cultural invention created in a specific time under specific life circumstances. In short, the abilities connected with counting and measuring originate from specific human capabilities of using language which has a recursive grammar, characterized by the hierarchic subordination of reality. The first rational invention in this field was the invention of the base, which became a constant element of rational numerical grammar. A certain paradox, based on the fact that the ability of such numerical grammar and vocabulary to develop was impossible during a large part of human history, is interpreted by Chomsky as a lack of ability to make use of the recursive nature of language. Following

---

[29] Grammar which is comprised of a finite collection of rules which make use of a finite supply of words is recursive as long as one may use it to create an infinite number of sentences. For in this case at least some of these rules have to be able to be used more than once in generating the same sentence. Hence, these rules are named recursive.
[30] Pollmann 2003, 28.

Chomsky's train of thought we can imagine that language recursiveness in the Palaeolithic was strongly attached to the narrative abilities of language communication, which means it was strongly anchored in group identity by storytelling[31]. When observing modern development from the historical perspective we can even say that the invention of the counting base, the development of mathematical concepts in the later periods may be understood as the beginning of the end of great narratives[32], or maybe even the beginning of the dehumanization of language, exemplified by computer language, which count only by means of the binary code.

Ethnological examples

In order to show the problem discussed here it is best to refer to ethnological examples. Ivanov presents some examples of Brazilian tribes, among which the concept of number can be expressed only by the use of special verbal constructions. For example, the Arara Indians, who belong to the same language group as the Pirahã, would say the following: *ma'wit ip #iy matet iagarokum–nem*, which means: "yesterday a man [in double number] caught [two] fish". Among the Kwaza, who come from the same group as the Arara, the numerals are mixed into nouns: *ka'nwã aky–'kai e'mã ele'le–tse*, which means "the car has more–than–two–wheels" (classifier)–three (=four wheels). This is of course very strange from our modern point of view, but it lets us assume that these examples illustrate the early method of expressing numbers. Moreover, users of languages described by Ivanov in fact have some problems with performing calculations and mathematical operations of an above–basic level, aiding this process with gestures. He even suggests that the development of mathematics has been much easier in languages in which numbers could be expressed as nouns just like objects, not as predicates as in the case of the Amazon languages he researched[33].

However, it is Benjamin Lee Whorf who became a classical figure in linguistic studies concerning native languages. This self–taught linguist, who attained a science degree in the field of chemistry, a researcher of the Hebrew and Aztec languages became renowned thanks to his theory concerning the influence of language on thinking. Two of its variations exist. The stronger version is based on the assumption that the language we use everyday determines our way of perceiving and understanding the

---

[31] Kordys 2006.
[32] Lyotard 1979.
[33] Ivanow 2007, 190.

world, which has been described by Whorf in detail in his studies of the Hopi language, which was published after his death[34]. In it he dealt with inter alia by means of expressing time and space in the Indian language. At present, the strong version does not find so many supporters as its weak version, which simply states that the structure of a given language influences the way of thinking about things and phenomena does not determine it completely. Nevertheless, the strong version has a couple of supporters.

One of them is Peter Gordon, who stated that language may influence the way people perceive spatial relations and the properties of objects. However, as Gordon points out, none of this research proves that the structure of our language is an impassable barrier; that it does not allow perceiving that which is seen by the use of other languages[35].

The Amazon tribes remain the key object of research concerning language determinism as not only do they mix concepts of number into their utterances in an odd way but also have a very limited numerical vocabulary. Hence, the Pirahã use the following vocabulary which concerns numbers: "hói" (dropping tone = "one") and "hoi' (rising tone = "two"). Greater numbers are referred to as "baagi' or "aibai", which simply means "a lot". This system may simply be called: one–two–a lot and, as researchers presume, a similar system could have been an initial understanding of number/measure in prehistory.

Referring to Whorf's theory, Gordon wondered if the Pirahã could bend their humble mathematical abilities in order to come closer to a real counting system even by making use of language's recursive feature. After performing the appropriate research he reached a conclusion that the Pirahã did not use recursiveness in creating new complex numerical expressions. They never used expressions such as "hoi–hoi" to describe greater numbers. Instead they used gesture language, which is the so called counting on fingers, as a technique which aids verbal counting. Gordon noticed that counting on fingers was often quite inaccurate even in the case of small numbers (up to 5). Moreover, the words "hói" and "hoi", which mean "one" and "two", were not used only to describe those numbers in particular. While the word "two" referred to a collection greater than the word "one", "one" was sometimes used to describe small numbers such as 2 or 3. In connection with this fact it seemed that the Pirahã language is unable to express precise mathematical numbers (table 1.1).

---

[34] Whorf 1956; 2002. The idea of linguistic determinism is called also the Sapir–Whorf hypothesis.
[35] Gordon 2004, 496–499.

Gordon claimed that societies which do not possess at least the simplest numerical system cannot differentiate sets comprised of more than 4 or 5 elements, which means they do not surpass the *subitizing* barrier which has been built in our brains through evolution. The language used by the Pirahã may be used to present and communicate in detail amounts that are less than 3 in number. Discussing this example of linguistic determinism, it is safe to say that the Pirahã's language is disproportionate to languages which have numerals, which allow the representation of precise mathematical numbers. Moreover, an interesting fact is that the Pirahã do not even possess a standard term for describing unity. Instead, the word "hói" bears the meaning of "roughly one" or "not many", which depends on the context. So does the strong version of Whorf's theory take place in this case? In fact, Gordon leans towards accounting this instance to the strong version of linguistic determinism. However, he adds that it is an exceptional case and it may even be the only one among ethnological languages[36].

Other research concerning the Pirahã language seem to only partly confirm Gordon's conclusions. This time the group of researchers focused on answering the question of whether the use of language which lacks numerals changes the way the speaker perceives the collections? Referring to Gordon's work, experiments have been conducted during which it occurred that, despite the lack of numerals, the studied natives were able to differentiate the precise sizes of the collections which contained even more than five elements, albeit with one exception. They were not able to memorize the numbers. The amount of wrong answers to the questions asked increased as time passed after the end of the experiment.

The results of the research clearly suggest that mathematical vocabulary is a cultural invention and not something universal for man. However, if words which describe numbers do not change the "external" representations of numbers and are only a type of cognitive technology which serves the ability to memorize, then perhaps something similar to mathematical precognition does exist?[37] Transforming it into a rational mathematical vocabulary would only be a matter of "extraction", as the mathematician Barrow proposes[38]. Nevertheless, the development of metrological–mathematical systems had to depend on social–economical circumstances. Societies which made intensive exchange contacts required new technologies of memorizing complex liabilities made during the exchange of goods. As the Pirahã did not perform intensive exchange, they

---

[36] Ibidem.
[37] Frank et al. 2008, 819–824.
[38] Barrow 1996.

did not develop the techniques required to do so and that is the reason why counting has such a marginal meaning for them.

The results of the second series of research did not confirm the Whorfian strong thesis in full, which as Gordon admitted could occur. According to them, language plays a comprehensive role allowing a sufficient codification of information concerning size, colour, spatial orientation, etc. However, in a situation when an appropriate code is underestimated or unnecessary, speakers communicate in a more descriptive, narrative, and gesticulative fashion–a way used by speakers of a language that does not possess appropriate vocabulary. Hence, colours, numbers and the vocabulary used to describe cardinal directions do not seem to change the cognitive processes which are their basis, contrary to what some researchers suggest, or if they do this is not made in a direct way. Instead, similarly as in the case of other technologies such as alphabetic scripture, the appropriate vocabulary grants its users the potential of a new–and more efficient–means of coding experiences[39].

Table. 1.1. The use of fingers and number words by the Pirahã (after Gordon 2004, reworked by the author)

Number of objects counted	Number words used	Number of fingers
1	hói (= 1)	?
2	hoí (= 2) aibaagi (= many)	2
3	hoí (= 2)	3
4	hoí (= 2) aibai (= many)	5 – 3
5	aibaagi (= many)	5
6	aibaagi (= many)	6 – 7
7	aibaagi (= many)	5 – 8
8	?	5 – 8 – 10
9	aibaagi (= many)	5 – 10
10		5

---

[39] Frank et al. 2008.

The problem of language determinism will probably engage researchers in the future. This matter will not be resolved in this book. Here we may only remark that the most recent studies concerning language and the development of mathematical cognition among the Pirahã tribes again seem to lean towards the initial thesis of Lee Whorf and Gordon. Caleb Everett's & Keren Madora's opinion is that the phenomena on a world scale upholds the idea that the ability to recognize precisely amounts that are greater than 3 is determined by a culturally developed concept, a mathematical vocabulary, which is not universal to all human societies[40].

Measuring or counting?

What is the difference between counting and measuring? From the cognitive point of view these are two different processes. According to Gregory Bateson, the number is not the same as quantity, as numbers are a product of counting, whereas quantity is a product of measuring. In the cognitive sense the concept of number must be manifested by the existence of a formula while the concept of quantity does not. Formulas are basic elements of the world, hence all living organisms are able to recognize them. The most obvious manifestations of formulas are simple geometric forms which can be expressed even by means of small numbers. That is why, according to Bateson, some animal species possess the ability to count small collections. The little time required by humans to recognize small numbers without counting is probably realized in this way. Bateson writes that numbers originate from the world of formulas, figures and numerical calculations and quantity belongs to the world of analogue and probabilistic calculations[41].

Let us employ a historical point of view. Is our perception of the passing of time a process of measuring or counting? Or maybe both?

The perception of astronomical changes, such as the change of daytime and seasons, the shape of the moon, etc. was probably the oldest method of measuring/counting. Natural cycles automatically become concepts which are used to describe changes in time. Also other phenomena, such as the growth of plants and animals, the change of the seasons as well as ageing and life in society (the chronology of one's experiences) were the earliest motive not to use number concepts to measure time. Archaeological data also suggests that the observation and adequate record

---

[40] Everett & Madora 2012, 130–141.
[41] Bateson 1996, 72.

of astronomical phenomena were the basis for conceptualizing time in all hunter–gatherer societies, at least from the middle Palaeolithic period, as it was then that mysterious cuts on bones appeared, which could have served as the first means of noting the passing of time[42]. It was indeed a childhood period of using mathematical concepts, although the word "mathematics" given here is somewhat exaggerated. Most probably some means of description, which are known from ethnological literature, were used then [43]. What could counting/measuring look like without mathematics?

Let us assume that we have to set the date of an important ritual. A shaman or other chosen specialist in recording dates, after performing some precise specified gestures, announces that the ritual shall be held on the thirteenth day of the eighth moon as from today. How could he express this period of time without using numbers, which he does not know? He could do it in the following way: "Many suns and many moons shall rise and set before the celebration comes. The moon which has just appeared must first grow and then shrink until it disappears completely. Then it must appear anew as many times as it is able from the little finger of my right hand to my right elbow. Then the sun must rise and set as many times as it can from the small finger of my right hand to my lips. Then the sun shall rise to indicate the Great Totem Ceremony."[44]

Counting and measuring by the use of one's own body as a special type of calculator is most probably the beginning of man's experience with mathematics, although still without the use of numbers. None of the body parts had any numerical meaning for our ancestors, it was only an element of the model which consisted of a specific and finite sequence of gestures. This meant that it was not enough to point to a given part of the body to specify the number or measure. A whole series of set gestures had to be made. In this early stage we may speak of the existence of a counting and measuring ritual, which was strictly connected with both astronomical observations as well as with the human body, which some archaeological discoveries, of which I write about later on, point to. Hence, the rituals connected with the human body and gesture are (apart from language) another potential source of mathematics.

Studies made in the field of brain research confirm these ethnological observations. The correspondence between counting and performing specific gestures (gesture numbers) is anchored in the distant past when the appropriate brain areas of the early *homo* species developed. Damage

---

[42] Marshack 1972.
[43] Ifrah 1985; 2006.
[44] Ifrah 2006, 77.

to these areas (the left parietal area) usually leads to an inability to recognize one's own fingers and to count, which has been called the Gerstman syndrome[45]. It seems that in regards to this the body as well as gestures made with it may be one of the oldest mathematical, or rather protomathematical, instruments. This fact is very important for this discussion, as on the following pages of this book I shall try to prove that the basis of the transformation of understanding mathematical rules in prehistoric times was concerned with moving the accent from the body onto the chosen elements of material culture.

Stages of the development of mathematics

Having the knowledge presented so far, let us try to make a draft model of the development of mathematic abilities in primeval times. Peter Damerow[46], whose main source of inspiration was Piaget's theory, has already drawn such a model and it will be described here briefly.

Stage 0. What came before mathematics?

There had to be a period in the history of mankind in which the concept of numbers, as well as developed mathematical abilities beyond that which Mother Nature equipped us with, did not exist. In that period even imprecise descriptions of amounts (compare above) were enough. Referring to Piaget's works Damerow named this the prearithmetic or premathematic stage. People could only perform simple quantifications of size and quantity of collections. They could only compare what was bigger and what was smaller or describe where there is a greater or smaller amount of objects. As for arithmetical operations, people were not able to use or describe them. No concept of number was known then and no numerical operations could have been performed.

Moreover, as Damerow points out, societies at this stage of development do not possess even the possibility to assess the size of collections by a one to one correspondence. The language of that time did not provide any means of expressing sequences of numbers even by the so called tallies–mathematical aids: fingers, other body parts, sticks or other objects. There were no words, symbols or signs that could be used to express mathematic concepts. The language then enabled man only to express the quantity of objects in a narrative way. Or in other words: the

---

[45] Ivanov 2007, 186–194.
[46] Damerow 1999.

amount aspect of the happenings and perceived objects was not different from the quantity aspect and the characteristic way of presenting them.

Where can we find such societies which lived on such a strange level of perceiving reality? These were probably the Palaeolithic societies, although perhaps we cannot treat all of them in the same fashion. The number of societies living today which make use of such a primitive understanding of numbers is very small, especially as most of them have experienced contact with European culture. These are the simplest hunter–gatherer societies known from ethnographic studies and descriptions: Australian aborigines, bushmen as well as some Amazon tribes, of which I mentioned earlier. However, this situation is changing dynamically as a result of the influence of western civilization and the number of people who remain at this level is rapidly decreasing.

The identification of the matters mentioned when speaking of Palaeolithic people is very difficult as we possess very little material evidence connected to them. Nevertheless, we can say that from this longest part of prehistory we do not have any direct evidence for the existence of mathematical abilities which exceed the premathematical level. To make this matter clear we must deliberate and focus on the issue of Palaeolithic signs, which are believed to have an arithmetic function[47].

On some bone, horn or stone artefacts from the Palaeolithic period we can find different types of cuts, from simple lines to the geometric sequences of symbols, which have been organized in rows. Moreover, mathematical concepts are believed to be seen in the wide spectrum of symbols which were left on walls of caves, in which Palaeolithic man lived[48]. Damerow provides a counterargument, however, according to which all Palaeolithic evidence lacks a repetitive sequence, which is characteristic for mathematical structures.

The above mentioned idea was researched by Alexander Marshack[49]. By examining the markings on Palaeolithic bones under a microscope he found out that they were made with different tools, and the process of making them was stretched over a significant period of time. The number of such cuts ranges from only a few to hundreds on a single object. A bone has been recorded to having 221 cuts[50]. Marschack underlines that there is no evidence to claim that these markings were seen in a numerical manner.

There are findings, however, which oppose such an unequivocal description of the Palaeolithic situation as completely devoid of

---

[47] Frolov 1974.
[48] Ibidem.
[49] Marshack 1972.
[50] Marshack 1972a, 817–828.

mathematical sense, as it would result from Damerow's and Marschak's definitions. On one of the bones from Dolní Věstonice a row of identical cuts parted with two longer cuts through the middle of the bone had been discovered. Grouped cuts were also found on other bones found there, which would suggest the existence of a counting base. Because of this finding some authors still claim that such types of symbols could be evidence of counting[51]. It would seem all the more probable that societies which inhabited the Moravian highlands at that time (that is 20,000–25,000 BC), among whom these symbols have been created, left behind them a very rich collection of archaeological resources, such as objects of refined art, complicated tools, richly equipped graves and sturdy dwellings. However, Marshack suggested that the markings on the Palaeolithic bones may have been connected with observing the moon and its phases, which means it could be a record in some sort of a lunar calendar. According to him, these notes express complexity, which suggests that they were made by singular (specialized?) individuals responsible for maintaining certain annual ceremonies and rituals[52]. Perhaps they used markings on bones instead of their own bodies to mark astronomical events. However, these notes were quite personal and did not have any numerical connotation, which is why they probably could be used by one person for one particular astronomical event or process. These markings would then be connected to the attributes of specific people, so they would not be free from their bodily control. No one else could interpret them properly. They could have been shamans, who due to their profession were highly interested in cosmological and mathematical phenomena. Marschak's main example, which points towards the astronomic meaning of Palaeolithic notation is a slate from Abri Blanchard which was covered by dozens of dimples, and interpreted by the researcher as a lunar calendar[53].

The concept of corporal links with early notations automatically comes to mind in the context of the Geißenklösterle cave findings. From this cave comes one of the oldest musical instruments–a bone flute, but our attention is focused on a small ivory slate ornamented on both sides and dated at 35,000 BC[54]. On one of its sides there is a picture of a man with his arms lifted up while on the other–dimple–shaped cuts made with a flint tool (Fig. 1.1). The edges of the slate have been cut at equal intervals. The most noticeable number is 13. It is repeated on the slate more than once,

---

[51] Kuckenburg 2004, 78–79.
[52] Marshack 1972a.
[53] See interpretation made by Rosenberg (2007).
[54] Hahn 1982, 1–12.

which according to Hansjürgen Müller–Beck [55] is a record from observations of the relationship of the Sun and the Moon (13 lunar months are equal to one solar year). This relationship was probably known to Palaeolithic hunters, whose nomadic way of life was subject to the basic changes in nature. Hence, during a year when the Sun returned to the same point on the horizon the Moon had already made 13 full cycles (four for each lunar phase from full moon to full moon). And as there are 4 rows of dimples, Müller–Beck connects them with the 4 lunar phases. He believes that it is hard to explain the finding differently.

What connection does this have with the human figure on the other side of the slate? Adorant which is the name that has been given to this famous picture, could have been strictly connected to the calendar notations described above. His figure has also been cut (mainly the arms). The presentation of the human body in the context of a calendar notation clearly suggests a connection with the method of measuring time as mentioned above.

Although it is hard for Müller–Beck to imagine another interpretation of the phenomenon from Geißenklösterle, archaeologist Michael Rappenglück presents a different one[56]. According to him the dimples' placing on the slate does suggest a time sequence, but slightly different. There are 88 dimples in all which roughly corresponds to the 3 synodical months (which is equal to 88.5 days) as well as the number of days during which the Betelgeuse star (α Ori) disappeared from view each year about 33,000 years ago while Orion appeared in its place. This shift took exactly nine months and was equal to the time of a human pregnancy. Hence, the nine month period during which Orion was visible in the sky marked the rule of thumb which determined the birth of a child after the harsh winter months. It was a rule set by the stars. According to Rappenglück, this hypothesis is even confirmed by the picture on the other side–the shape of the anthropomorphic figure (Adorant) resembles the Orion constellation. In other words, the ivory slate from Geißenklösterle would be an example of a specialized tool for controlling the human pregnancy period in relation to the movement of the Orion constellation and the Moon.

At the moment it is not our task to decide which interpretation is more probable. It is enough that we remember their shared basis, which is emphasizing the relations between the human body and measuring/counting the passing of time, which is reflected on a small slate from Swabian Jura.

---

[55] Müller–Beck 2001, 66–68.
[56] Rappenglück 2012.

Fig. 1.1 Adorant from the Geißenklösterle cave
(drawing by Dobrawa Jaracz).

Stage 1. Protoarithmetical

It was the first introduction of the concept of number. At this stage the basic arithmetical equations as a series of techniques of categorizing and size controlling different objects were established. We may surmise that these primeval mathematical concepts concerned only some chosen aspects of reality. In other words, not everything was counted and measured except for the objects most important from the perspective of a given culture.

We may connect the protoarithmetical stage with the moment the first material representations of quantity in a one to one correspondence appeared in human societies. The symbolic representation of quantity corresponding to counted objects and objects of different shapes or symbols began to appear. In literature they were referred to as *tallie*s or *calculi*. They could be knots, pellets, cones, small stones, sticks and so on. These objects could afterwards be organized into structures which corresponded to the objects in question.

At this point it should be underlined that Damerow's concept proposes a quite strict caesura. When combining the development of the Palaeolithic and Neolithic cultures he notices the change in quality, although not one which would qualify to a new stage. According to him the Neolithic revolution did not change the fundamentals of the cognitive development already achieved in the earlier period. The change in quality was introduced along with the urban revolution in the Near East[57].

Damerow places the Palaeolithic symbols and cuts on the same level as the Neolithic tokens, which is not completely correct. Although we shall discuss Near Eastern tokens in the next chapter, at this point we must underline that the essential difference between the two was that the Palaeolithic notation systems were "personal" and did not function in changeable social relations. Tokens were quite the opposite–they were made to function in interpersonal relations and were meant for specific economic contexts (chapter 2). Moreover, Damerow treats Denise Schmandt–Besserat's results concerning the development of tokens in the Near East quite selectively stating that earlier forms were not used for the purpose of simple administration. This statement stands in conflict with the results of the researcher, as she clearly pointed out that early tokens were used for administrative tasks[58].

The reader will see that on the following pages of this book I will try to revise the theory of the development of mathematical abilities proposed by Damerow. The majority of our gathered knowledge concerning the beginnings of mathematical concepts comes from the Near East where the basic elements of that process, which in effect led to the creation of writing, have been reconstructed. However, in prehistoric Europe the means of number presentation developed, which were much different from the ones known from archaeological literature. In Europe, especially in its northern parts, scripture and marks of a centrally managed palace economy have not been found, although this does not mean that other means of understanding mathematical concepts did not develop here. It seems we are faced with a different path, not with its lack of development. The aim of this book is to prove that this stage is quite well observed in archaeological material, it only requires an adequate theoretical point of view.

Stage 2. Symbol–based arithmetic

Symbol-based arithmetic was created when cognitive constructions connected with the size and amount of objects were supported by second (or higher) order external representations. It then became possible to

---

[57] Damerow 1996.
[58] Schmandt–Besserat 1982, 871–878; 1992; 2007.

manipulate reality by the use of symbols without the physical involvement of the objects which they referred to. This stage corresponds to the creation of administrative scriptures on early clay slates in the city–states of Mesopotamia when the concept of number had separated counted objects (compare chapter 2).

Stage 3. Arithmetic

Arithmetic as a stage was reached as a result of the introduction of the coding of both concepts of number as well as the operations used to manipulate them in written language. Symbol–based arithmetic originated from the earlier stage. Its characteristic element is the creation of a complex collection of numerical symbols and formal rules for applying them. Initially symbols were used for counting metrological units, although unlike in the earlier stage they were supplied with strict rules of transformation for performing operations on numerical signs. A formalization of sums occurred. Even if sign transformations which were part of the framework of these rules correspond to the meaning of the signs themselves they remained stable, as operations performed with their use were not bound to the changing conditions of managing the symbolic units of real objects.

This is a brief presentation of the prehistory of mathematics. During this process the direct relationship between the numerical unit and the symbol which represented it became looser. Today it has reached such a level of abstraction that it is a separate resource of knowledge, which does not require contact with the material objects that are used.

Body, mind, metaphors

Our initial point for further discussion should be the previously outlined border between the ability to count and not having that ability. Humans, like some other animal species (chimpanzees, rats, parrots, dogs, pigeons) are born with a mathematical potential. Rats and chimpanzees, if not taught by the use of special techniques, remain at the same level throughout their whole life. However, the situation is similar in the case of humans, which yet again emphasizes the assumption that mathematical abilities are in a way a cultural invention. This clear discord between the quick perception of small collections and counting, which is something more, has been called *subitizing*. There are two qualities which probably designate the border between the mathematical perception of reality and something completely opposite[59].

True arithmetic requires abilities which go far beyond the cursory assessment of quantity or perceiving small collections. Beyond this border lay the abilities of precise counting, which are typical for man. George Lakoff and Rafael Núñez decided to distribute these abilities into component parts. It appears that such a simple action requires the use of many additional cognitive mechanisms that were unknown before. Hence, besides the requirement of possessing a language of description (symbolic language), counting presents us with the following requirements:

    1. The ability to group objects: to count anything we must first group the appropriate objects visually, mentally and tangibly.

    2. The ability to organize: our body is a natural model of order, especially fingers and toes. Other objects in nature rarely show this type of organization and must be organized "artificially", the easiest method to achieve this is associating them to specific fingers. That is why the most common method of counting observed by ethnologists among simple societies was counting on fingers, although this is not a universal rule.

    3. The ability to make pairs: it is required to determine the correspondence each time between the counted objects and body parts.

    4. The ability to remember: in order to count efficiently we must remember which finger or body part has been associated already, and at the same time which objects have already been counted.

    5. The ability to perceive borders: we must be able to determine when the objects to be counted have ended.

    6. Associating the cardinal number: the last number during the counting procedure is the ordinal number. It has to be replaced with a cardinal number which defines the size of the counted group of objects. This size no longer has an order meaning.

    7. The ability of independent organizing: when counting we must be aware that the cardinal number, which defines the size of the group, is independent of the order in which the objects are counted. This ability allows us to see that, regardless from the order, the size of the group always remains the same[60].

This is not all. We need the abilities listed above when we count to 4 (it must be emphasized that we *count*, not perceive a size of such value without numerically organizing and naming it). However, if we want to continue counting we need additional cognitive abilities:

---

[59] Trick 1992, 257–300.
[60] Lakoff & Núñez 2000, 51.

8. The ability to group combinatorially: it is a cognitive mechanism which allows combining a greater group of objects (real or imaginative) into one collection.
9. The ability to symbolize: it allows to recognize symbols (physical objects or words) with numbers, which are conceptual entities.

This is still not all! Counting is only the beginning of arithmetic–the foundations of mathematics. To understand arithmetical operations we need additional cognitive resources:
10. The ability to metaphorize: we need the ability to conceptualize cardinal numbers and arithmetical operations in the framework of human experiences, e.g. with grouping objects, perceiving the structure of parts and whole objects, with experience of distance, movement and localization, etc.
11. The ability to mix concepts: creating a correspondence between idea areas distant from each other as well as combining them into one common metaphor. Lakoff and Núñez call this a metaphoric blend[61].

The ability to create complex metaphors, which is to mix different spheres of meaning (e.g. such spheres as social relations with natural science or technology) is, according to Lakoff and Núñez, a basic mechanism enabling people to tackle mathematics. This concept seems to harmonize with Stephen Mithen's conclusions, that cognitive fluidity, which is the ability to easily combine content from different spheres of human activity: the social, craft and environmental spheres (information concerning the natural habitat), evolved among *Homo sapiens* between 100,000 and 35,000 years ago. Meanwhile, the minds of our earlier ancestors, which functioned on the basis of the Swiss–army–knife mentality, were more specialized, which meant that combining technical and social content was difficult. The symbolization and metaphorization of different spheres of meaning was impossible at that time[62].

Lakoff and Núñez claim that the metaphors concerning simple everyday activities, such as gathering objects in piles, moving containers with objects inside them, enable the understanding of arithmetic operations. Abstract mathematical thinking becomes available to us with the help of metaphoric blends and because of them we can understand numbers as e.g. points on a straight line. Let us consider a couple of basic metaphors, by which we are able to see our world in a mathematical way.

---

[61] Ibidem, 52.
[62] Mithen 1996, 205.

Object collection metaphor

In connection with Piaget's research, Lakoff and Núñez confirm that manipulating objects becomes the basis of understanding mathematical rules in early childhood. When a child sees three blocks he/she will automatically connect the three blocks into one group. If we take one block from a child then it will understand that the group of blocks was reduced to two. This type of basic experience and correlation in a child's life lead to creating neural connections in the brain, which is a consequence of gaining experience while manipulating objects (taking away and adding objects to the collection).

However, dealing with objects also leads to understanding the rules of manipulating them which suggests that adding is automatically connected with something greater while subtracting with something smaller. Words such as "big" or "small", which metaphorically mean sizes of objects or collections of objects, are also used to mark numbers, for example in questions we often ask children during their mathematical education, such as "what is bigger, 5 or 7?" or statements such as "two is smaller than four". This metaphor is based so deep in our minds that we often have to think twice to understand that numbers are not physical objects and do not have a size[63]. Hence, the basis of the object collection metaphor is another metaphor which identifies numbers with objects.

The metaphor of collecting objects is comprised of the following elements:
 a) the numbers are like collections of comparable objects
 b) the size of the collection is like the numbers' size
 c) adding numbers is like joining collections
 d) taking an object away is like removing the smaller collection from the bigger one
 e) the smallest collection is 1

It seems that the object collection metaphor could only develop in specific socio-economical conditions. Although they could potentially be used by Palaeolithic societies which, according to Mithen's hypothesis, achieved the cognitive fluidity stadium, it seems equally correct that this was fully possible only when the objects, filled with rich symbolism (mostly graphic) of the surrounding nature retreated into the background, were replaced by a derived collection of crops enclosed in vessels and the domesticated animals in farming societies[64]. These animals automatically

---

[63] Lakoff & Núñez 2000, 56.
[64] Wierciński 2004.

became a derived collection–a collection, property, which management and manipulation in complex human relations began to assume a greater meaning. The constant control of these, as well as their exchange among the intensifying and dynamically changing social relations became the basis of Neolithic management. The information that someone owes someone two pigs or a cow, whilst being someone else's creditor not only creates the knowledge of that person's biography but most importantly a potential for the development of condensed description possibilities (through number). Of course the same concerns crops.

Let us examine another example, Caves in the Palaeolithic period could not be perceived as elements of land, as they were occupied or treated as sanctuaries filled with deep symbolism. We do not find cave paintings in all of them but the fact that humans lived in them, burying the dead in them, emphasizes the special relationship between humans and caves. They were a central object of cosmologic beliefs, probably playing the role of a connection between worlds: the upper and lower, which is a characteristic element of shamanic beliefs[65]. Caves possess a complicated network of corridors descending to the underworld. In other words, the perception of caves and mathematic abstractness does not go in pairs. This situation is different among societies which had abandoned caves. They know of their existence but they avoid them and only mix them into local myths and narratives (e.g. caves as birthplaces of gods). It is then that such an element of culture (stories) cease to be filled with live content and becomes a topos, an island in a sea of similar narratives. Hence, if caves became perceived only from the outside, in exchange for building anthropogenic constructions, to which we shall return in chapter 3, a step towards understanding them as a point in space has been made.

Zero

The object collection metaphor generates a certain mathematical problem. What happens when we remove seven objects from a collection of seven objects? The result of such a manipulation is hard to understand as a collection. Nothing remains, hence the collection ceases to exist; it is gone. If we want such an operation to remain a number we must change our way of thinking and understand the lack of a collection as an empty collection. A collection which has no objects. In order to do this it is

---

[65] Clottes & Lewis–Williams 1996; 2009.

necessary to compose an additional conceptual metaphor which created something from nothing[66].

The measuring stick metaphor

The oldest way of measuring small lengths and constructions was the measuring stick. It was known both in the Near East and in ancient Egypt[67]. In the simplest cases the measuring stick was a reflection of body parts which were used as measures of length (hands, fingers, ankles, and feet).

A measuring stick may be seen as a physical object even if it is only an imaginary fragment of space. However, as Lakoff and Núñez point out, this object is also unidimensional as it can be infinite. The problem of the measuring stick's infiniteness, before the development of philosophy, had to be examined on a religious/ritual level, which I shall elaborate in chapter 3. In the abstract version the metaphor of the measuring stick corresponds with fragments in Euclid's geometry. As a result, this metaphor has the special status of a physical link of the object (the physical measuring stick) with numbers which define its size. The measuring stick metaphor carries close conceptual metaphors of arithmetic as a movement along a line, which allows us to see numbers as far or as close to each other. On the margins of our discussion we may add that the consequence of using the measuring stick metaphor in its Euclidean version was the discovery of irrational numbers[68].

The conceptualization of the measuring stick metaphor seems to be as fundamental for man's prehistory as the object collection metaphor or zero. However, this metaphor has a special meaning for prehistoric Europe, as I will attempt to prove in this book, or at least it found an exceptionally original reflection there. It seems that unlike the prehistoric Near East, where the development of mathematical concepts was based on the object collection metaphor, the measuring stick played a major role in Europe. This metaphor was used initially to execute the concept of number in constructing Neolithic buildings: megalithic stone constructions and maybe even the houses of the first farmers. After that it spread its influence onto (or rather came into) human relations, where macrolithic tools appeared. I think that probably **the unique character of the European Eneolithic, represented by the urge for large tools,**

---

[66] Lakoff & Núñez 2000, 64.
[67] Kubba 1998, 91–103.
[68] Lakoff & Núñez 2000, 71.

corresponded to the protonumerical fantasy and the metaphorization of the measuring stick phenomenon which is the early conceptualization of measure/number in the physical dimension of length, which I will attempt to prove in the following chapters.

Among the Eneolithic as well as later societies this metaphor was subject to further abstraction thanks to metallurgy technology, which requires a new conceptual approach. The stick had been replaced with the weight of metal, which is the length number for a weight number, which is the basis of real mathematics. It is most probably then that number abstractness appeared in Europe for the first time. It seems that the difference between Europe and the Near East is reflected in the different distribution of accents in the process of forming the first mathematical concepts. In Europe the introduction of the first mathematical concepts appeared from the metaphorization of the measuring stick, while in the Near East the main focus was put on the object collection metaphor. This difference will be mentioned in the rest of the book and summed up in chapter 6.

Body, mind, and space

According to Lakoff and Núñez, a mind closed in a body, or an embodied mind, is the basis for conceptualizing reality. The body, the structure of our brain and our everyday earthly movement, manipulating our body, all of these are the soul of human reasoning including, as the above mentioned authors stress, mathematical reasoning. The meaning of mathematical symbols does not lie in symbols alone, or in the way they can be manipulated by the use of rules. Neither does it lie in their interpretation as theoretical models or axioms, which remain without interpretation. All in all, mathematical thinking is a derivative of everyday thinking and is a part of embodied cognition[69].

The concept of the embodied perception of the world comes from the philosophy of Maurice Merleau-Ponty, the heir of Heidegger's and Husserl's ideas, which he called the perception phenomenology. According to this concept, the subject must be seen as connected with perceived objects and things. In reference to other people, who would also be perceived as objects, the metaphor of a society in which the subject becomes its part, may be created in this way. The subject is an embodiment of the object – man is implicated in things first of all by his

---

[69] Lakoff & Núñez 2000, 49.

body[70]. The objects surrounding us are inseparably connected to man as a subject, thus becoming an extension of man. This means that we broaden our existence in the world through time and space by objects[71].

The phenomenon of perception was an essential part of epistemology, although the Cartesian division of man into a body and a thinking subject introduced a considerable distance to the world, which was connected with the existence of artificial categories and copies of the external world. After giving it more thought, the Cartesian thinking subject seems to be an element of the body–machine, quite loosely connected with it. Merleau–Ponty opposed this vision by writing: "*my body is the fabric into which all objects are woven, and it is, at least in relation to the perceived world, the general instrument of my comprehension*"[72]. Hence, man, as he stated, existed only through his own body, whereas that body creates a synthesis, it is a mind–body, a psychophysical being. The body becomes an incarnation of the subject, present at all times and impossible to discard both in the process of perception as well as the reflection of it. My body is my anchor in life, as wrote Merleau–Ponty, and "*my main measure in my relation to the world–time, space and its content is the universal thing...the universal measurement*"[73]. We may in fact make this theory which has deep metaphoric content one of the main theses of this work.

Lakoff and Núñez relate to Merleau–Ponty's theory exactly at this point pointing out that by creating and understanding abstract ideas by the use of more concrete concepts as well as bodily existence in the world people mainly make use of cognitive mechanisms which result from rendering reality through a sensor–motor system. Hence, the way we reach certain abstract concepts is based on the fact that we perceive the things in our world as changing, moving, shifting shapes in response to the way we act on them with our bodies. The mechanism, during which abstract ideas are described by the use of concrete objects and events Lakoff names the conceptual metaphor, which is based on our everyday experience [74]. Conceptual metaphors definitely take part in creating the basis of mathematics, for example when we perceive numbers as points laid in a straight line[75].

In the light of the above mentioned arguments, Lakoff and Núñez give four daring statements in their book, which are as follows:

---

[70] Merleau–Ponty 1945; 2001; 2002.
[71] Jensen 2000, 53–67.
[72] Merleau–Ponty 2002, 273.
[73] Jensen 2000, 58.
[74] Lakoff & Johnson 2003, 272.
[75] Lakoff & Núñez 2000, 5.

1. As human beings we do not have access to the world of Plato's transcendental mathematical ideas. Faith in them is not a matter of science.
2. The only mathematics that man is capable of discovering is the mathematics of his own mind limited and organized by brain work.
3. The analysis of mathematical ideas shows that human mind mathematics uses conceptual metaphors in creating mathematic knowledge.
4. Hence, mathematics cannot be a part of transcendental Platonic mathematics, if such exists.

Lakoff and Núñez are not the only ones who try to point out the inadequacy of cognitive theories which separate the human mind and the material world. The concept of an embodied mind is faced with a composition in the forms of the "situated mind", "extended mind", "enacted mind", "distributed mind" or "mediated mind" [76]. Neurobiologists also relate to it through experiments [77]. Although differently named and accented, all the mentioned concepts are also linked by the question present throughout this book: where does the mind end and the rest of the world start? Or rather: where does the mind end and the material world start? If we refer to the problem of early mathematics to this we may also ask the following question: was it created only in our heads or did the material world play, and is still playing, a role in this process?

It is not my intention to discuss each of the concepts mentioned in detail. They all share a common feature, which is questioning mechanical division into the mind, the body and the material world. That is all we need. We may say that if human perception (cognition) is generally a means of engagement in the world, then material culture is in fact not only a product of our mind but a part of it. The relation between the outer world and human perception (cognition) is not a certain abstract representation or some outer form of acting at a distance but a relation of ontological indivisibility, as Malafouris writes. Although the cognitive base of the mind is the creation of the mental representations of all things in the world, it is not necessary. For we can think through things, in action, without the need for creating mental representations[78]. This is undoubtedly a non–mechanical vision of the mind.

---

[76] Malafouris 2004, 57.
[77] Damasio 1994; 1999.
[78] Malafouris 2004, 58.

Number and space in brain research

Studies of the correspondence between spatial perception and number has provided some interesting results. These studies seem to confirm Lakoff and Núñez's assumptions of the elemental abilities of the human mind which concern the conceptualization of linear measure as they illustrate the direct link between space and understanding numerical relationships. In other words, if the perception of number and spatial measure are strictly connected to each other, the cognitive condition of our brain clearly implies the development of geometry. However, it is important that these studies allow us to support the hypothesis according to which linear measure was one of the first levels of understanding numbers. Thanks to these properties of the human mind we know that the extended artefact could gain a numerical aspect in the early stages of cultural development, which is in fact what I state in chapter 4.

Several simple experiments have been performed on volunteers which resulted in the identification of SNARC: spatial numeric association response code. During these experiments it occurred that the volunteers found answers concerning greater numbers faster if they are placed in the right area of vision. The results showed that smaller numbers seem to be a mirror reflection: if they are placed on the left people identify them faster. Moreover, it has been observed that numbers are able to somewhat "bend", directing attention towards the right or left side, depending on whether people focus on greater or smaller numbers. Human eyes are subject to a specific deviation when they have to point to the centre of a line, which consists of the word: twotwotwotwotwotwo and the word: nineninenineninenine. In the first case the trials of pointing to the centre of the line had a tendency to deviate to the left, while in the second–as it is easy to predict–towards the right[79].

However, it is important to mention that in the research mentioned above the cultural aspect became statistically relevant. Researchers assume that it is probably connected with the literary tradition from which the researchers subjects came. And so, people who came from countries with a right to left writing order had opposite connotations connected with number and space. For them, greater numbers were connected with the left side, not the right as in the case of Europeans. Because this research was not conducted on illiterate people we do not know what the deviation factor would be of perceiving left or right.

---

[79] Malafouris 2010, 35–42.

Conclusions: the role of material culture

In the conclusions of this chapter let us organize the facts. As far as the language–mathematics relations are concerned there are two basic mindsets. The followers of the Whorfian mindset believe that language determines the human thinking of reality (and mathematical thinking at the same time) and states that possessing words for describing numbers, as well as their various relations, is an essential condition for the development of mathematics. The protagonists of this mindset eagerly call upon anthropomorphic research concerning some small social groups, among which there are no numerals and which cannot perform any mathematical operations. This is the case of such groups as e.g. African bushmen, Australian aborigines as well as some Amazon Indian tribes.

However, opponents of this mindset present arguments which give a much more complex image of the problem. Among other things they point out that there are some areas in the brain connected with performing arithmetic operations, which are far from those areas of the brain that are responsible for language. Moreover, observations made in the field of some brain injuries, which deprive people of mathematical abilities do not correspond with a decrease of language abilities and vice versa. For example, observations made on autistic patients, who have difficulty in expressing themselves, often do not confirm the decrease of their mathematical abilities. These observations show that calculations can be made without activating speech abilities, which is the reason why some researchers suggest the existence of a numerical sense[80].

There are experiments which support the thesis according to which the precise determining of quantity is created not only by language alone. Language allows only for a comparison of meaning, very efficiently coding the information concerning quantity in the same way it does in relation to other features such as colour or cardinal directions (it is better to say "we go north" or: "this blouse is red" than "we go in the direction where the moss covers the tree" or "this blouse is the colour of my cheeks when I get angry"). However, in a situation when the appropriate code is silenced or needless, people act in a similar way to those who use language without proper numerals. Researchers used to state that in such a case, in some specific socio–economical conditions, numbers are simply not so crucial, hence they are not paid attention to so much in everyday life. In addition, besides the lack of numerals, the lack of a counting routine is a difficulty in itself. If there is no need to count anything...

---

[80] Malafouris 2010, 38–39.

The science laboratory references mentioned above seem to suggest that although the possession of words and concepts for describing mathematical laws in a language may be helpful for the development of mathematics, it is not necessary. In that case, should we assume that language only eased the development of mathematics, but was not necessary for it? Should we return to Plato?

At this point we must remember that counting may have many forms, one of them and probably the easiest being counting by using one's body, most often with fingers. However, this does not mean that fingers have a numerical value, they are only used as tokens, an aid for recording events. In such a case, the lack of numerals is replaced by the body. And so the expression such as "the tenth day of the second Moon" may be presented in the following fashion: "the Moon, right ankle, day, left arm", as I presented above. In the same way we may perform simple calculations and, in the case of greater quantities, involve a whole local society for this purpose[81]. However, because this is a very limited system, going beyond the human body is required to make use of mathematics more efficiently.

It appears that neither language nor our corporeality is the only means of understanding the development of mathematics in prehistoric times. The sphere of material production, which helped the mathematical development process, was until present a somewhat unappreciated and not properly discussed sphere of human activity. This problem will be discussed in the forthcoming chapters. The thesis stating that the development of mathematics would be impossible without the intensification of production, which is generally present in the Neolithic period[82], is even more interesting if we are aware of the above statement. The relationship between man and his person made creations become so complicated that it was necessary to create a new means of expression, a new language and vocabulary which would describe the relationship between man and objects, including the metaphoric view of the world of objects[83]. Tokens, bone marks, poles, etc., as both material objects and media used to manipulate other objects had a very significant part in the history of mathematics as they allowed the first concepts of number to develop, the processes which Walter Ong dubbed discretization. Fingers used for counting as well as other tokens remain units connected with the body called *digits*. Counting on fingers is still definitely anthropomorphic, as they are separated at the ends but remain a part of the body at their

---

[81] Ifrah 2006, 72–83.
[82] Wierciński 2004, 34.
[83] Tilley 1999.

base[84]. However, the introduction of tokens and other material media detached the *digits* and enabled the real penetration of social relations, in other words their interpersonalization. These digits were undoubtedly material metaphors still connected with the body, ritual, power and control, helping to shape the relationship between ownership and object. They entwined with social relations, penetrated them leading to the further development of discretization, and at the same time the increase of abstraction.

To summarize this chapter we may state four things. Firstly, we know that mathematical abilities have their basis in both biological structure and the culture of man. They have a direct connection to our corporeal presence in the world and how we manipulate the world with our bodies. They are also connected to us being able to represent the world verbally by means of symbols and metaphors in human communication. Last but not least, they are connected with the fact that we have been producing objects which have been important for us for a very long time. I will try to apply this knowledge in the following pages with regards to specific archaeological cases.

---

[84] Ong 2009, 219.

Chapter Two

Ex Oriente Lux or How Numbers Were Invented

In the previous chapter, I proposed that both inborn predispositions as well as linguistic abilities might not have been enough to initiate such intellectual development that was responsible for the mathematical understanding of the world. The consequence of such a presumption is going beyond the complex language–brain–body relation (or language–body) and introducing the influence of material culture into the analysis. Basically, the best solution would be to have an adequate example to which we could refer. Fortunately, it just so happens that we do. It is the prehistory of the Near East, in which three basic stages of development of the concept of number have been identified. Let us have a closer look into this history in the present chapter.

Clay tokens

In many Neolithic Near Eastern settlements, researchers have found small clay and, less often stone, objects which had regular geometric forms. These were most often in the shape of a sphere, disc, cylinder or cone. Initially these objects were not given much attention, interpreting them as toys or pawns, until Leo Oppenheim, an American researcher, found an interesting lead in 1958 when an egg–shaped, hollow clay container from the ancient city of Nuzi in northern Iraq caught his attention. Eight lines of cuneiform script text had been inscribed on its surface which spoke of "stones for sheep and goats", where 48 pieces of animals from several species were counted in total. Although the container was empty, some shrewd archaeologist or curator added a note to the container which stated that when it was found it contained 48 stones which were afterwards lost. Oppenheim immediately guessed that these 48 stones could be related to the 48 animals counted in the text and reached the appropriate conclusion that this was probably a previously unknown ancient method of record and administration[1].

During the search for further evidence of this administrative system Oppenheim came across a collection of cuneiform texts in the palace in Nuzi which in relation to the animal count consequently mentioned "depositing", "removing" and "moving" stones. Examples of these are: "these sheep are in the possession of (the name of the owner), appropriate stones have not been deposited"; or "a sheep belonging to (name), the corresponding stone has not been removed". Oppenheim concluded that a systematic record of the ruler's possessions had been kept in the palace by keeping an amount of stones which represented living animals in containers kept for that purpose. They were organized into such categories as the animal's gender, age and species. In such a simple way the possession of any goods (we may imagine that it was used not only for animals) could be under the constant control of any appointed administrators[2].

Over twenty years later archaeologist Denise Schmandt–Besserat took up the task of presenting a holistic explanation of this strange administrative system. While searching through museum resources of Near Eastern prehistory she came across some small geometrically–shaped objects, even among very early collections of artefacts connected with the pre–pottery Neolithic period. The oldest collections were not accompanied by any additional pottery vessels, they were found "loose". They had been found among collections of artefacts from Neolithic settlements in Iraq, Iran, Syria, Turkey, and Israel. Before Schmandt–Besserat's work, tokens were still considered as charms or pawns (despite Oppenheim's suggestions). Schmandt–Besserat dubbed these tokens (Fig. 2.1).

Fig. 2.1 Early Neolithic tokens from the Near East
(drawing by Dobrawa Jaracz).

---

[1] Kuckenburg 2004, 135.
[2] Ibidem.

Schmandt–Besserat has shown that these items have occurred even at some early pre–pottery Neolithic sites geographically connected with areas of wild crops. The oldest come from between 9,000 BC and 8,000 BC at five sites in Syria and Iran and soon after they were discovered in quite a widespread area over almost all of the Near East. Perhaps at the very beginning there were only 10 basic shapes which came in different sizes. These shapes survived in an almost unchanged state for almost 4,000 years as they were still used in 4,000 BC.

Today the explanation for the invention of the first tokens in the Near East between 9,000 BC and 8,000 BC is widely accepted as the introduction of the first counting/measuring systems as well as systems for recording the available resources in the process of converting to farming. Naturally there are voices of criticism to which we shall return later on[3]. These systems met the needs of calculating and memorizing a greater number of tasks and resources connected with farming activity and household economy. It is also expected that this new phenomenon did not remain without some connection with the developing social inequity structures.

Along with the development of farming, settlements became larger. In its early stages, the Mureybet settlement was mostly centred on a hunter–gatherer economy (9,000 BC), no tokens have so far been found. At that time the settlement covered an area of approximately half a hectare. However, in its third stage (8,000 BC), in which it grew to a village covering about 3 hectares, the first tokens were found[4]. A settlement of that size was inhabited by a more varied society than the hunter–gatherer groups. Hence, the correlation of these three phenomena: the introduction of systematic farming, the settlement's growth, which pointed to greater social diversity, and the introduction of tokens allow us to connect them into a joint complex. In other Near Eastern Neolithic sites similar relationships have been noted. In Tepe Asiab, Ganj Dareh, Tell Aswad, and Cheikh Hassan the appearance of tokens was always connected with traces of farming, more specifically the cultivation of crops[5].

The Near East region of the Neolithic period has been the main object of interest for prehistorians for a long time as it is the area of the initial development of farming. Equally current and interesting are the questions concerning the development of the first Near Eastern civilizations, factors such as deep social inequities, the development of the palace economy, the elite and first court culture. It seems that from 4,000 BC Mesopotamia had

---

[3] Michalowski 1993, 996–999.
[4] Schmandt–Besserat 1992; 2007.
[5] Ibidem.

been an area of important demographic, technical and cultural transformation which approximately 700 years later led to the creation of the historical city–states of Sumer and Akkad. However, there was the matter of the environment (the so called irrigational theory), which put stress on the organization of field labour, could not explain all elements of this complex process. The starting point of urbanization was the rapid population increase, evidence being the explosive growth of small and large villages surrounding some central sites such as Uruk. This demographic explosion had been initiated by three features: natural increase of settled populations, gradual settling of nomadic tribes and the influx of people, probably coming from the northern areas of Mesopotamia. New arrivals settled near trade centres and places of worship[6]. They were attracted by the technical developments of that period, such as the invention of the coulter, a four wheeled cart, as well as the sail[7]. Intensive development of the potter's wheel and metallurgy took place. A large part of the population started to specialize in different activities, such as trade and exchange, craft (production of common and luxury goods), and thus a new type of management was required. Hence, a communication system which used tokens, taking different forms, was on the increase.

Of course, it is difficult to clearly define the influence of tokens into this process but today we may say without doubt that the Near Eastern rationalized communication systems were also a socio–technical medium which contained information concerning resolutions, agreements between groups and individual partners. They took part in far–distance exchange between administrative centres and distant manufacturing sites[8]. In this way they became a potential means of control and verification, which was quickly adapted by the first rulers as a basis for the authoritarian rule of individuals over the production potential of the society.

Use context

The archaeological context tells us much of the tokens' purpose. Many of them have been found inside buildings located in a central square, which suggests their specific role in the settlements. A classic example of this is the Tell Abada site from 6,000 BC located in northern Iraq. A clear correspondence between the larger buildings and their diverse equipment

---

[6] Roux 1985; 2003.
[7] Roux 2003, 65.
[8] Oates 1993, 403–422.

was noted. Wealth was equated with architecture. In the longest building located in the centre of the settlement an interesting assemblage of evidence was found: special burial rights, a large concentration of stone tools, decorated slates, stone clubs, but most importantly administrative tools, that is tokens as well as administrative proto–tablets. The last two items were found only in this particular building[9]. These circumstances have been confirmed to have been continued in the subsequent three settlement stages, so researchers suppose that the administrative functions could have been inherited.

Of course the situation was not so clear in all of the settlements. In fact, as Schmandt–Besserat writes, in most of the early settlements tokens were found in different places, seemingly without any context. In many sites they were discovered in layers between constructions, sometimes along with animal bones and broken pottery, which means they were left along with rubbish. According to her, different goods were recorded in the framework of small communes during celebrations and ceremonies when tokens were probably thrown away, hence removing the record itself to which they referred[10].

We need to be aware of the very long development line of jettons which are present in a vast part of the Near Eastern area regardless of the subsequent cultural changes in the societies which came and went during this process. This fact has been used as an accusation towards Shmandt–Besserat's interpretation[11]. However, this accusation is irrefutable as this is an identical case like many other inventions which break culture barriers. The wheeled cart had been implemented by many cultural groups by 4,000 BC[12]. A similar situation occurred in the case of macrolithic industries, which is analysed in chapter 4. The characteristic feature of macrolithic industries was inter alia that they were utilized by different societies throughout a long period of time. It is known that the great inventions of mankind are able to rise up over locality.

Field observations show that old methods of recording co–existed with the new ones for a long time. The consequence and evolution of early information record systems described by Schmandt–Besserat have been confirmed archeologically at the Suza site in Iran. They were discovered there in a clear stratigraphic sequence. Tokens and pottery vessels have been found in the lower strata, whereas the upper strata revealed more and

---

[9] Stein 1994, 35–46.
[10] Schmandt–Besserat 2007, 28.
[11] Michalowski 1993.
[12] Pigott 1983; Kruk & Milisauskas 1999, 162–170.

more complex recording techniques on the first slates covered with proto–script and cylindrical stamps, which represented administrative activities.

Some basic shapes which did not change throughout the Neolithic period can be distinguished. These were spheres, cones, discs and cylinders (Fig. 2.1). The first meaningful revolution reflected in the jetton system can be dated from 4,000 BC, along with the intensified urban development of the southern regions of Mesopotamia. Archaeologists who have specialized in the cultures of that region often speak of an urban revolution which was based on the intensive growth of the size of some settlements known to us from the first cuneiform texts. Many new local variants of tokens appear in that period and the whole system has hundreds of symbols. According to Schmandt–Besserat, there is a strict connection between the system of complex symbols and the creation of the first cities–nations. In all of the more important cities of the ancient Near East, such as Uruk, Susa, Chogha Mish and Habuba Kabira, complex tokens have been found in archaeological strata which also contained cylindrical stamps being evidence of the presence of a controlling elite, and characteristic vessels, which are represented on clay slates as measures of grain[13]. These vessels were used to measure food portions for labourers carrying out temple maintenance duties[14].

New methods, which made use of the greater diversity of marked tokens, created possibilities for a more precise settling of liabilities brought to a temple or a palace for the developed bureaucracy of 4,000 BC. They described more details, such as the species, age and gender of the animals. For example, they could differentiate between sheep, rams and lambs, while simple cylinders only informed about the number of animals. Complex tokens could record the amount of wheat or barley, while simple cones and spheres only showed the general amount of grain. The increasing variety of shapes and markings of tokens is evidence of the increasing diversification of counted and administered objects. Processed food products (bread, olives, and poultry), crafted products (wool, clothing, cloth, mats, rope, furniture and tools) as well as luxury goods (perfumes and metal objects) were all collected as part of the taxes. Products such as bread, olives, cloth, pots, axes and perfumes had surely been known for a long time. However, their being recorded was new[15].

In Uruk, tokens have been found in the most impressive buildings of the 5th strata, such as the Stone–Cone Temple and the Limestone Temple, while in the complex of buildings F, G, H in the Eanna district a group of

---

[13] Nissen 1988, 84.
[14] Damerow & Englund 1987, 117–166.
[15] Schmandt–Besserat 2007, 103.

75 tokens were discovered *in situ*, although Schmandt–Besserat does not give a more precise description of the discoveries. Tokens in the temple context have also been found in Suza in areas which researchers interpreted as warehouses and workshops. It was sometimes stated literally that tokens were found in granaries. There they found a lot of pots with elongated nozzles, still filled with substances with unknown origins. In Habuba Kabira tokens were also found in the vicinity of the most impressive buildings. Clay stamps, elongated bullas, envelopes and slates with stamped symbols were found as well as flint tools, pottery vessels and weaver weights. A distinct artefact also found in all of these sites was the characteristic clay pot–shaped or beaker–shaped vessels that have a flourished inlet.

There were a small amount of tokens found in the graves. And if so, these were most often made of stone and not in clay. Hence, we can assume that we are faced with special symbolic equipment, as Schmandt–Besserat suggests. The graves that were discovered containing tokens are quite early, perhaps later on, the custom of placing objects into these graves was abandoned. In them we find the remains of people who are described on early slates as "calculating masters" or "stones people"[16].

An interesting matter worth discussing in the context of the development of economic and administrative methods is the increasing dichotomy of gender. David Wengrow noticed its first signs which were characterized from varied archaeological evidence[17]. For example, in Tell Sabi Abyat among strata dated to 6,200 BC a division of the site into two types of buildings had been noted: round tholos tomb–type constructions and rectangular constructions. These two types seem clearly correlated to a specific economic activity assigned to gender. Items used in textile and food production, such as pestles, weaver weights, spindle whorls and bone needles were found in round constructions, whereas items such as grain vessels, tokens and cylindrical stamps, which were a form of acknowledging economic identity were found in the rectangular constructions. In other rectangular buildings in this site anthropomorphic figures, which were systematically broken around the neck and thigh, were discovered. This discovery is a very characteristic feature of Neolithic culture from the Near East to Central Europe and is interpreted as one of the ways of cultivating social communication[18].

According to Wengrow, the final stage of the development mentioned above was the trisection of the Near Eastern household, expressed also

---

[16] Ifrah 2006, 364; Schmandt–Besserat 2007, 98.
[17] Wengrow 1998, 783–95.
[18] Biehl 2003; Becker 2008, 119–127; Hansen 2001, 93–106.

symbolically as one of the metaphors which organized social life. The trisected construction became a basic structure for domestic life and as the farm metaphor spread into the sphere of administration, production and ritual activity, this influenced the further facilitation of new ethics as well as an increase in productivity[19]. This took place in the Ubaid period between 5,000 BC and 4,300 BC, directly before the dynamic development of urbanization and the explosive development of tokens.

Tokens and counting

Different shapes of tokens used to record different goods are a sign of concrete counting, the most archaic technique which lasted for a very long period and is even used to this day in far corners of the world[20]. What does such a system look like? It is quite hard to imagine. Concrete counting is characterized by using different sets of numerals, so–called digits, in order to count different things. On the language level this system was probably quite limited, which was why it was supported by tokens. In reality, some relics of the system have survived to this day in the form of everyday expressions, when we say: twins, triplets, quadruplets or solo, duet, trio, quartet. So the word triplets represents the number 3 as well as three new born children to the same mother, although they cannot be separated and used to describe something else. A quartet is a way to describe a musical band comprised of four musicians. As Barrow, a mathematician, suggests, we only need to acknowledge that this was the way counting was performed in the beginning[21]. Similarly, in the case of tokens, one oval token meant one jug of olive oil, not giving the possibility of separating the jug and the oil. As these two pieces of information could not be separated from each other, the value was expressed in a one to one exchange.

Near Eastern tokens are extremely important for determining the evolution of mathematical thinking as they show the whole progress. It is from their example that we may trace the history of the development of mathematical abstractness. This schemata may be presented as follows:

1. 8,000 BC–3,500 BC. During this long period of time tokens were used in concrete counting systems, which meant that this mechanism was strongly embedded in the human psyche and society. Each category of objects required a special type of token, e.g. ovoids represented jugs of

---

[19] Wengrow 1998, 790.
[20] Diakonof 1983, 78–98.
[21] Barrow 1999, 70.

olive oil, while cones–measures of grain. When determining numerals in a "one to one" relationship, that is three ovoids = three jugs of olive oil. In this period the concept of number did not exist without a connection to the object. We may perceive tokens as a metaphorization of counting on fingers, the basis of the forming of a 10 counting base in the Near East. Still they did not represent numbers.

2. 4,000 BC–3,100 BC. Marks imprinted on envelopes and the first clay slates were still used in a "one to one" relationship and this is evidence of a strong connection between an object and the concept of number. Hence, concrete counting dominated. In the same period the existence of two methods of using tokens has been noted. The first one was based on enclosing them in clay envelopes (bullae). The second was threading them on a string like bead necklaces. Both methods were used to identify efficiently the transactions or arrangements which were additionally confirmed by a special stamp imprinted on an envelope or added to a string with tokens.

3. It is probably correct to say that in about 3,100 BC in the ancient city of Uruk digits were invented. The division of the concept of number depending on the type of the objects counted probably occurred among specialized administrators/accountants. Each of the concepts was represented by a different sign; the type of goods was represented by pictogram engravings and the number of units was represented by symbols, which were digits. For the first time symbols began to represent numbers in an abstract way. Digits were created as a result of a change in the meaning of symbols which were used earlier for marking measures of grain and animals–most often by the use of tokens. Hence, this was an innovation based on making a conceptual turn in understanding mathematical rules. The fact that this innovation was created among specialists may mean that the rest of society was intellectually separated from this development.

4. 3,100 BC–2,500 BC. In this period the archaic methods of counting were diminishing in some areas. This is evidence that the transition from concrete counting to abstract counting lasted for several centuries[22].

The greatest contribution by tokens was the promotion of abstract thinking. Replacing material goods with small standardized clay objects was the fundamental rule of this system. As a result, exchange could be performed more rationally as tokens led to abstracting goods from their actual context. Objects, which were previously perceived through a strict

---

[22] Schmandt–Besserat 2007, 118.

relation with the natural environment, became objects of abstract manipulation in human relations. Animals and plants could be controlled in a completely new way–through placing symbols that related to them into specific columns.

Moreover, tokens led to the abstract perception of time as they allowed administrators to manage goods which were detached from context, regardless of whether they were in the field, the temple pen or whether they even physically existed. By the use of tokens, administrators could perform simple arithmetic calculations (adding, subtracting, and dividing). Managing data in the form of tables, columns, promoted abstractness, since a ceremony could have been planned by managing virtual data[23]. It is beyond doubt that the system of tokens extended more and more in the Near East from 8,000 BC which led people (although probably not all) to acquire new competences. The internalization of the these abilities over thousands of years opened up an area in the human mind for using more and more abstract concepts, such as mathematical rules and script[24].

Early metrology

Research into the early notation systems in the Near East gave grounds for the reconstruction of counting and measuring methods, which were completely different from those known today. The analyses of numerical symbols used in archaic texts gave us a series of extremely interesting results concerning the history of mathematics and the development of the ways of perceiving mathematical and metrological relations. Early in the 1970s Aisik Wajman, a Russian sumerologist, and Jöran Friberg, a Swedish mathematician, stated that early numerical symbols were probably used in different meanings and were part of diverse systems of numbers and measures. In 1987 another pair of researchers: Peter Damerow and Robert K. Englund, supplemented and polished the results of this research into a coherent system[25].

Damerow and Englund distinguished about 60 numerical symbols which they segregated into 5 basic systems of number and measures. These could be further divided into 15 sub–systems, each being used in a separate discipline: for determining the amount of grain by means of measurement capacity and for measuring areas arable land etc. The numerical symbols used in these systems were, in part, differently shaped,

---

[23] Schmandt–Besserat 2010, 33.
[24] Schmandt–Besserat 2009, 145–154.
[25] Damerow & Englund 1987, 117–166; Nissen et al. 1993.

and as such almost impossible to be mistaken. However, sometimes identical symbols were used for different values. Damerow and Englund concluded that pre–metrical counting and measuring systems were not based on an abstract concept of number, but in a related order of metrological–numeric units–with their place in the ranking.

Imprinted shapes, which were the successors of clay tokens, arranged themselves into an organized ranking and did not have a value independent from the context. This resulted in the existence of many different capacity measuring systems, as well as many length measuring systems (different measures were used for buildings, land and materials such as cloth). Similar systems were used in historic Europe[26]. That is why symbols that are present on slates are dubbed numerical symbols, which only possess a relational value, unlike numbers which possessed an abstract numerical value. For example the metrological system of grains was composed of six different symbols set in a 5–6–10–3–10 relation. The small dot marked as "N–14" in this system would represent a value connected with its place in a sequence between "N–1" and "N–45". Its numerical value was not assigned to a symbol. It is clear when comparing it with other known area measuring systems, which were composed of numerical symbols organized in the following relation: 6–10–3–6–10. Hence, the numeric value of the "N–14" symbol would mean 180 in an area measurement system, not 30 as in the case of the grain measurement system. These systems did not operate independent values but values dependant on what was counted and measured. The numerical values placed sometimes under the symbols of Near Eastern measuring systems could have been a mere technical aid for today's readers (Fig. 2.2; 2.3).

The number 1 was represented by a short wedge, which also represented a small measure of grain. The numbers 2, 3, 4, 5 etc. were represented by an appropriate number of wedges. Number 10 was marked with a circle symbol which was also interpreted as a large measure of grain. As we are faced with a 60 system, the number 60 is represented by a large wedge, 600–a large stippled wedge and 3,600–a large circle symbol. These examples show that numerical symbols made on the first slates retained their basic meaning of grain measures while gaining an additional meaning–digits. Hence, in many early texts symbols representing measures of grain present in early texts are the same as digits which represent their amount. The number of grain rations given to labourers were recorded with the same symbols as the number of labourers. It is beyond doubt that this system requires concentration and attention and

---

[26] Kula 1986; 2004.

could be managed only by a trained specialist. Nevertheless, the introduction of digits had shortened the process of recording. For example 33 jugs of olive oil required only 6 symbols: three circles and three wedges, not imprinting 33 symbols one by one[27].

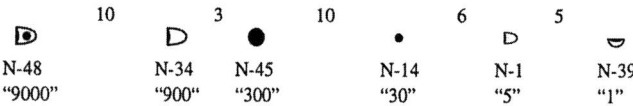

Fig. 2.2 Protoliterate grain measurement system recorded on Mesopotamian tablets (after Justus 1999, reworked by the author).

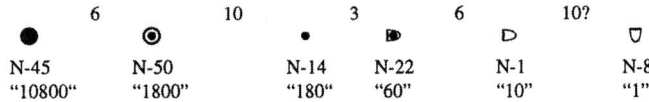

Fig. 2.3 Protoliterate area measurement system recorded on Mesopotamian tablets (after Justus 1999, reworked by the author).

How were units of measure created?

When taking into consideration the above mentioned arguments we may say that cones, spheres or discs in prehistoric times represented measures of grain on a protomathematical level. They symbolized vessels in which grain was kept, e.g. a small basket, a large basket or a granary. These types of measurements could be compared to modern descriptions, such as a glass of sugar or a pint of beer. These were not yet units of measure in the numerical sense and only after some time, along with the increase of administration and central managing, were the appropriate standards introduced. A pint of beer can certainly be a numerical measure– it only needs to become a norm which may be treated as a unit. Then half a pint of beer or two pints will become further units in the metrological system. However, the transition from one state to another seems quite subtle and certainly difficult to capture in archaeological material. Schmandt–Besserat suggests that the evolution of measures and weights should be seen as the following:

1. In the beginning everyday objects, such as vessels of grain, a bucket of flour or a jug of water, existed without any numerical connotation.

---

[27] Schmandt–Besserat 2007, 114.

2. As a result of the development of managing resources, a certain standardization of these units occurred. The size of the vessels became more or less similar and the royal foot was chosen as the basic unit of measuring length.

3. Different units of the same goods present on a greater area had been subjected to a common scale. One large basket = 10 small baskets or one foot = 12 inches[28].

The Near Eastern tokens present from 8,000 BC till the second half of 4,000 BC seem to correspond with the first stage when traditional everyday vessels without their own numerical value determined measure. Nothing suggests that any steps had been taken to standardize these vessels in that period.

It was only the rolled edge beaker, which appeared in the Near East in the Uruk period (3,500 BC–3,000 BC) that may point to the first processes in the standardization of the capacity to measure. According to the first slates containing protoscript, these neat vessels were used for retail grain distribution (most likely cereal) for public labourers and were manufactured in two basic sizes[29]. Schmandt–Besserat suggests that the calibration of measurement units and the creating of a scale of their multiples did not occur before 3,100–3,000 BC, which she bases on a slate from the Jebel Aruda site in Syria dated at approximately 3,200 BC. This slate presents 3 large symbols next to which there are 22 round symbols, which should be interpreted as a record of 3 large measures of grain and 22 medium measures of grain. Hence, the slate shows that each unit of measure was still counted separately. The abstracting of the metrological system was only developed in 3,000 BC and was unknown earlier[30]. It is also important to remember that the standardization of weights and measures was slowly becoming visible in the Near East. In 1,000 BC larger cities such as Babylon and Assur possessed their own units of measure. One of the greatest achievements of King Darius was the weights and measures reform, which introduced greater homogeneity in the whole of the Persian empire.

All of this points to the fact that this evolution of measures and numbers was strongly connected with the production and use of small tokens. In the first stage of this evolution vessels were not yet perceived in the numerical sense and their proto–numerical representations were tokens. They played the role of a medium in human communication while

---

[28] Schmandt–Besserat 2010, 27–34.
[29] Nissen 1988; Nissen et al. 1993.
[30] Schmandt–Besserat 2007, 115.

progressively shaping the rudiments of mathematical perception from the object–token relation. The existence of objects as well as tokens would not be enough for the development of mathematical concepts. It was an indispensable tandem. Moreover, by developing our analysis we can reach the conclusion that the circulation of tokens in social relations helped the development of perceiving everyday objects in an abstract way. In time, vessels were not only seen as objects associated with a narrative content but also as containers; the circulation of object representation, instead of the objects themselves, which developed specialized abstract thinking among the people responsible for gathering and manipulating this data. Hence, in the early stage of such manipulations some possibilities of specialization became available in the field of managing and a separation from intellectual development and other social levels.

The beginnings of the system of counting

If we remember from chapter 1 there are examples of societies among which only the concept of "one", "two" and "many" exist, which can be found among ethnographic notes while studying the phenomenon of mathematics. The traces of such a simple approach to the problem of quantity may still be found among modern languages. John Barrow, a mathematician, points them out among European languages, in which special expressions, such as *primo, secundo, and tertio* are present. This would mean, according to him, that the words which define ordinal numbers are much older than other numerals[31].

The counting system in which only words which describe the two first numbers is called a binary system and is the simplest one possible. Let us compare it with the decimal system when numbers from 1 to 10 have their own distinct words. After these the counting process starts somewhat anew. Why cannot we "extend" the binary system in a similar way? Possessing only two numerals is the main problem of doing so, since such a method of counting is too limited and does not allow us to express greater numbers. Moreover, according to Barrow, counting by the use of two numerals is not proper counting but only perceiving quantity as a pair, a pair and a single unit and two pairs[32]. He claims that societies which are not able to count to more than five should not be regarded as ones being able to count at all. They only know adjectives which describe some states in which objects appear. These societies often employ conjugations and

---

[31] Barrow 1999, 69.
[32] Ibidem, 96.

declinations in the plural, double, triple and even quadruple form, which is the case of the Amazonian languages described in the previous chapter. They perceive the doubleness or trebleness of an object.

The methods of counting by using parts of the human body are closer to abstract counting, as Barrow claims, as they are a sign of possessing a counting base (e.g. five fingers or ten fingers). By the use of this system we may count many things, although it is not very practical and it burdens the memory. We could choose a certain value, e.g. one hand, as a certain constant to which we could repeatedly return after counting each five. In this way a counting base is formed, which eventually becomes the basis of all real counting systems. Today most people count by using a 10 base, even although this was not always the case. Other counting bases existed in the past: 4, 5, 10, 20, and 60.

According to Barrow, the effectiveness of memory counting begins with using a 5 base. This could be how the beginning of more developed systems, such as the decimal, vigesimal and 60 systems, looked like. An interesting fact is that traces of the strange vigesimal system can be found even in Europe. It could have been very widespread in prehistoric times in areas such as Spain, the British Isles, France and Scandinavia. Evidence of this may be the modern word describing 80 in French, *"quatre–vingts"*, which means "four twenties". In parchments from the Middle Ages the number 80 was written as IIII^{xx}, which is 4 x 20, while the number 133 was written as $Vi^{xx}XIII$, which is 6 x 20 + 13. Similar expressions may be found in Old–English[33].

Moreover, Barrow claims that the Sumerians also possessed a vigesimal system but they expanded it to a sexagesimal one. Perhaps the development of a central administration and exchange between 4,000 and 3,000 BC in that region had a direct influence on this. Their odd system was based on counting from 5, 10, and 20 to 50 in the following way:

1, 2, 3, 4, 5, 5 + 1, 5 + 2, 5 + 3, 5 + 4, 10, ..., 20, ..., 10 x 3, ..., 20 x 2, 20 x 2 +10

Each number between 10 and 60 had its own name. While the number 60 was described as a new unit (*gesh*) in the same way as one. Larger numbers were written as multiples or powers of 60 in the following way:

600 (*gesh–u* = 60 x 10)
3,600 (*shar* = 60 x 60)

---

[33] Ibidem, 107.

36,000 (*shar–u* = 60 x 60 x 10)
216,000 (*shar–gal* = 60 x 60 x 60)
2,160,000 (*shar–gal–u* = 60 x 60 x 60 x 10)
12,960,000 (*shar–gal–shu–nu–tag* = 60 x 60 x 60 x 60)

The number 60 had a special meaning in ancient Mesopotamia cultures. Most researchers assume that it is connected with the metrological division which could have been developed in the Neolithic period in the Near East, where the smallest units had a value of 1/60. In this way it was easy to define precisely the value of goods as fractions were avoided.

Schmandt–Besserat assumes that traces of a very primitive counting system which was characteristic for its inability to exceed the number three, still remain in the Sumerian language. The main numeral three (*eš*) is identical to the plural form morpheme, which she interprets as the remains of a period in which people were only able to count to three, and "three" also meant "a lot" or "many"[34], similar to the Pirahã language described in chapter 1. Moreover, the existence of several Sumerian methods of counting based on a ternary system points to the fact that abstract counting in the Near East was preceded by an archaic and concrete one. Texts show examples of the existence of inter alia and methods of counting such as the following:

1. *merga* = one
2. *taka* = two
3. *peš* = three
4. *pešbala* = more than three
5. *pešbalage* = one more than three
6. *pešbalagege* = one–one more than three
7. *pešpešge* = three–three–one

Schmandt–Besserat assumes that the threshold of counting above three has been crossed, thanks to innovations connected with the development of farming, new methods of storing goods and a redistributing economy. According to her, the diversity of tokens is evidence that, despite the fact that the first farmers worked out the concept of a closed collection, they still used concrete counting. In other words they possessed no concept of number, existing independently from measures of grain or quantity of animals, which could be used for all types of goods. The presence of

---

[34] Schmandt–Besserat 2007, 107.

tokens confirms the hypothesis that in the prehistoric Near East concrete counting came earlier than abstract counting[35].

Damerow agrees with Schmandt–Besserat's interpretation and adds that the words used for describing numbers in the Sumerian language still bear the marks of a sexagesimal structure, although their record comes from a later period of that culture's development. The sexagesimal system corresponds with counting dependant of the context of more than a concrete one. He claims that only about 2,000 BC did an abstract notation system, which was independent from the counted objects, develop among a group of Babylonian priests. It was then that the possibilities for the development of real mathematics were created by means of symbols and formal rules alone[36].

A very interesting fact concerning the evolution of the Near Eastern counting system is that it seems to be devoid of religious inclinations. All of the information we presently possess concerning the early Sumerian culture suggests that the early use of numbers and mathematical vocabulary was limited only to practical operations. This fact stands in clear opposition to the image of the developed stages of ancient cultures, among which numerology used by priests had been a common phenomenon[37]. This change is connected with the development of the priesthood which was interested in facilitating its own position in society. This is a very important observation as it confirms the validity of the thesis, according to which the introduction of the concept of number into social relations meant the characteristic rationalization of these relations. In the first stage concepts of number led to the verbalization of sacrum, which allowed the development of the secularized side of the argumentation in the framework of social communication[38]. It was only in the next stage that they were used again in the ritual context. The development of mathematical thinking in prehistory should be based on two levels: on the cognitive development level and the social level, which was connected with the intentional actions made by individuals. As a result of the use of numerical and metrological messages, completely new possibilities of distinguishing a person from the general society appeared. It appears that the number is particularly useful at underlining the human ego and individuality, whose first level was connected with going beyond the local group. Following this trace we arrive at the thesis of the expert and media visionary Marshal McLuchan, who stated that counting was

---

[35] Ibidem, 111.
[36] Damerow 1999.
[37] Barrow 1999, 113.
[38] Habermas 1981; 1999.

breaking the unity of a tribe. Numbers, and afterwards phonetic letters, were the first means of "human fragmentization" and stripping him/her from social affiliation[39].

What about Europe?

The Neolithic settlers arrived in Europe from the Near East. Apart from the many other material adoptions from its area, the European Neolithic also received tokens, of which considerable information has been given above, as heritage. Did these small objects play the same role in Europe as they did in their own nation?

The appearance of European tokens is limited to South–East Europe, areas culturally connected with the direct colonization of Anatolia[40], or with the later influence from that direction[41]. Both the material and spiritual cultures of that region still clearly represent the Anatolian initial pattern, although it is certain that newcomers had to react dynamically each time to the changing environmental conditions. They had to experiment with new crops or animals, which is why the spread of farming could not proceed mechanically[42]. As a result, the Neolithic tribe's people of the temperate area of Europe began to differ greatly from that of Greece or the Near East.

The history of studying European tokens is similar to the Near Eastern one. They were long seen as objects of an unknown function and described by such enigmatic terms as stamps, clay cones, miniature clay objects, buttons, toys, etc. They are not numerous in the most important stages of the Neolithic period. Only about a dozen pieces have been found in the proto–scripture stage at such sites as Argissa, Nesosis, Sesklo and Pyrasos. They appear some time later, e.g. in strata dated 6,800–5,800 BC in Thessaly, where they can be identified with the sesklo culture. At this stage the development of settlements occurs, houses become larger, even megaron–shaped constructions with two or three rooms appear. The size of the settlements increased to 2 hectares. At first sight it seems that the introduction of tokens in South–Eastern Europe was connected with the spread of the Neolithic people. However, they did not appear in the stage of the spread of the Neolithic culture; at the stage of experimenting in the new environment, but only during the stage of consolidation and rise.

---

[39] McLuhan 1964, 161.
[40] Renfrew 1990; 2001, 216.
[41] Videiko 2009, 179–186.
[42] Bellwood 2005, 72.

Perhaps it was at this time that contacts with the homeland of farming intensified and thus changed their character. According to Mihael Budja, a disproportionate large number of tokens were found at Nea Nikomedea (about 6,000 BC), a site connected with white and red painted pottery. This site is perceived as a transmitter of culture trends from Anatolia[43].

The typological analogies of European tokens with their Near Eastern counterparts also create some difficulties. In Europe it is not easy to distinguish the early stage of simple and complex tokens. It is quite the opposite–tokens of that type appear much later than in the Near East. Despite this, Budja is willing to count the small artefacts found at such sites as Argissa, Suophli, Magula, Achilleion, Sesklo, Genticoi and Vrbica as vessel–shaped tokens, which are the equivalents of complex tokens of the Near East. These would be much older than their Near Eastern counterparts. This is why they have a particular meaning for Budja. They appear in such places as the sites in Dalmatia and at Vrbica in Croatia, which are connected with cardium pottery. They were also found in tombs which, according to him, are also interesting with regards to parallels with the Near East. Schmandt–Besserat states that tokens most often appeared in ostentatious tombs or those which were distinct in some other way[44].

In the next chronological cut in Southern Europe we are faced with the appearance of new types of tokens, among which cylindrical forms are the most distinct. Some of them have been described in literature as cylindrical stamps, which according to Budja was a mistake. These tokens appeared over a wide area: in Turkey, in the Carpathian Basin, Moldova and Thessaly (Fig. 2.4).

The analysis of the geographical spread of these small objects in Southern Europe presents us with an interesting image. The distribution of cones and cylinders (cylinder stamps) shows a tendency towards excluding each other, overlaying themselves only in the western Balkans. Southern Europe is characterized by the particularly high presence of tokens as well as other objects which bore a communication function, such as clay stamps. Whereas in the western Balkans the distribution of cones is connected with a particularly large amount of small zoomorphic figures. Budja assumes that it is not a coincidence and he sees this as evidence for the existence of a common communication network in which these small objects were engaged[45].

Despite the many difficulties connected with interpreting these mysterious objects, the Slovenian archaeologist demonstrates how South

---

[43] Budja 2003, 115–130.
[44] Schmandt–Besserat 2007, 33–34.
[45] Budja 1999, 219–235.

European tokens, apart from their significant formal differences from their Near Eastern prototypes, function in the network of interregional communication in south–eastern Europe and suggests that we should see signs of far more individualized and distinct dynamics of cultural changes in this part of the continent. According to him, some other types of jetton would have been developed there, which could hardly be compared to the Near Eastern specimens

Unfortunately, the number of systematical studies of European tokens is insufficient. Budja goes far in his interpretations, suggesting a reinterpretation of many objects. He proposed inter alia, to treat the items labelled so far as "zoomorphic charms" as tokens, for they were an effect of contacts with local hunting societies, which took part in creating the cultural image in Europe. Hence, the basic problem of European tokens was a significant formal distinction from their Near Eastern counterparts. Moreover, their number, when compared to their initial Neolithic area, is not high. So far a few dozen have been found while thousands were discovered at the Near Eastern sites. However, this does not mean that these objects can be ignored.

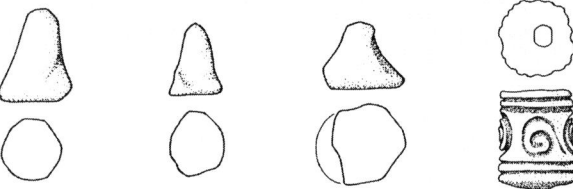

Fig. 2.4 European Neolithic tokens (after Budja 1999).

The matter of European tokens seems to be more interesting in the context of the Balkan scriptures problem, which had been developing in the same areas as tokens. This research was seen in a different light after the discovery of the Tartaria slates, which seemed identical to the oldest Near Eastern administrative slates. However, it appears that these artefacts are much older and come from the end of 4,000 BC; therefore they predate any known scriptures by approximately 1,500 years! There are other differences as well.

According to Marco Merlini, the "scriptures" of the Danube basin is an example of a very archaic writing system and probably could not encompass the full pronunciation of long, complex utterances. These phonetic elements, which would reflect the phonic complexity of human speech, have not been recognized in it so far. It is only composed of logograms, ideograms and pictograms, the last one being the fewest. Hence, these signs would have a much clearer connection with the sphere of concept and idea than human speech in the Saussurean *langue* sense. However, Merlini puts the "Balkan scriptures" in the same group as the Indus Valley scriptures, early–Elamite and with Olmec glyphs. According to him, these scriptures were somewhat *in statu nascendi*, developing alongside several other means of communication of that period, such as religious symbols, mnemonic signs, astrological and ritual signs, numeric notations, family or social identifiers or even the signs of ownership present on some artefacts. The knowledge that such different communication subsystems could actually function among the Neolithic and Eneolithic Balkan societies provides researchers with serious interpretational difficulties. In such a situation the specific signs or expressions of the scripture could coexist on the same object along with other informational signs. It is very probable that this was the case, although it is not possible to differentiate these matters[46].

Harald Haarmann also compares them to the Indus Valley scriptures, saying that the content of abstract symbols in these systems was similar. The number of symbols does not exceed the known norms, it reaches slightly more than 1,000 symbols and is comparable with other early script systems. Kuckenburg has a different view on this matter, as he estimates the number of symbols in this system to be only slightly above 200, which would not qualify it as script system[47]. Early notation systems in Mesopotamia operated with more than 700 symbols, twice as many as have been noted in Chinese bone scriptures, which developed into a very complex script system. Egyptian civilization scripture contained from 700 to 1,000 hieroglyphic symbols. Moreover, Haarmann calls upon other archaeological evidence which would point towards the existence of a complex notation and calendar system in the "Danube basin civilization"; calendars imprinted on pottery vessels being examples of this[48].

However, according to the archaeologist Adrej Starovič some archaeological evidence exists which supports the idea of rationalization which appeared in the framework of those symbolic notations. In the latter stages of Neolithic societies, signs and symbols had been imprinted on pottery objects only after burning. They were scraped in a special style on vessels, which Starovič explains as the using of signs for practical purposes, perhaps in the context of managing the household, as many numerical systems could be identified on these vessels. Who could have

---

[46] Merlini 2006, 233–251; 2009.
[47] Kuckenburg 2004, 123.
[48] Haarmann 2006, 221–232; Durman 2001, 215–226.

been the author and recipient of these short messages? Because of the growing import of goods such as high quality flint, from which macrolithic blades were made, copper, salt is also noted in the Neolithic period, we may assume that the development of intensive exchange, during which manufacturers–specialists or travelling merchants introduced innovations aimed at rationalizing the symbolic message[49].

Nevertheless, most of the signs discovered on pottery objects in the region of the so called Danubian civilization are symbolic signs of a narrative character which are hard to interpret explicitly. Although symbols which could be counted as numerical notations have been discovered, they remain a minor part of the findings. Their meaning should be perceived far from purely administrative functions, which were from the very beginning the basis for the development of Near Eastern tokens and slates. Most researchers correctly interpret them as ritual signs, carrying a large amount of narrative content[50]. It is the basic difference from the Near Eastern development, during which narrative contents have been successfully isolated in the token production context. The same has not been observed in the Balkans.

Conclusions

As Schmandt–Besserat indicates, the first tokens which appeared in the Near East are related to the production of food, specifically growing cereal crops. In Tell Murajbant these tokens were found only in the layers with a large amount of grain silos, which were not present in earlier layers where only evidence of hunting and collecting wild cereal existed. The Murajbat settlement spread rapidly and its population grew. Moreover, cereal grain was the basis of the everyday diet in such settlements as Gandż Dare Tepe, Tepe Asjab, Tell Aswad and Szajch Hasan, whereas no signs of bone shards, which would indicate the domestication of animals, were found there. Thus, tokens were apparently connected with farming.

The context of European discoveries is different and deviates from that of the ones made in the Near East. Budja points to three examples (Nea Nikomedea, Rakitovo, and Donja Branjevina) in which European tokens appeared in almost identical sets: anthropomorphic figures, small clay altars and zoomorphic plates[51]. This may suggest that European counting tokens were strongly connected to serving ceremonial functions. Chapman

---

[49] Starovič 2006, 253–260.
[50] Winn 2009, 49–62.
[51] Budja 2003, 115–130.

probably would wish to share this interpretation. In his book about the process of fragmentation of objects in Europe he states that the fragmentation of anthropomorphic figures was related to the practice of binding human relations (including relations with ancestors) by material means–the tokens of significance–such as inter alia anthropomorphic figures and their fragments[52]. In a modern linguistic interpretation, the tokens found in Europe seem to have a broader and less specialized meaning than those in the Near East. Perhaps they were rather tokens of significance than tokens of number. Moreover, the south European tokens seem to share the same epistemological status as the so called "Balkan Script". When analyzing both of these items we can observe a significant difference between Europe and the Near East. In the Balkans, both tokens and writing can be distinguished as obvious elements of religious ceremonies. Meanwhile, those in the Near East have an administrative character. The difference between the Balkan and Near Eastern tokens is the same as between Balkan and Near Eastern writing.

Near Eastern tokens seem to be more closely connected to crops. The matter of the rapid growth of early settlements, stimulated by cultivating crops and new communication methods in this area seems coherent. The pressure on the environment emphasized the role of crop records in the intensifying social relations which led tokens to become the basic model of early rationalized communication systems in the Near East. They were introduced in Europe along with the farming model into a different environment, in which contact with hunting societies played an important role. In this case it seems obvious that we should not expect the same path of development as a more archaic communication of a narrative character could be a better bond for these contacts[53].

Let us think of the consequences of this development. Near Eastern tokens were a completely new means of communication as well as a new medium. The conceptual revolution, in the case of how the tokens were used, was based on what special meaning was attached to each specific form – such as a cone, sphere or disc – unlike symbols appearing on Palaeolithic bones (the discussion of its true purpose raises a lot of controversy) and which gave rise to an infinite number of possible interpretations. Each clay token was a significant symbol on its own with one specific meaning. This meaning was connected to the context which was represented by its form. Symbols on Palaeolithic bones, when detached from any time–space context in which they were created, did not

---

[52] Chapmann 2000, 75.
[53] Dzbyński 2008.

convey any message, whilst the language of tokens was understood by any member of the society in question. The system allows the simultaneous manipulation of information concerning many objects, which means managing information in a complex and rational way which was unknown before. For the first time in history the possibility of containing large amounts of information concerning any number of different objects without depending on human memory existed. New tokens of a set shape could be manufactured at almost any moment.

This revolutionary medium had its disadvantages, however. Real, three–dimensional and tangible tokens had one basic flaw: it was difficult, if not impossible, to operate a larger amount of data. It appears from the content of the envelopes that they always contained only a small number of tokens, which meant that only a small amount of information was passed on each time. The envelopes themselves were so small that many objects could not fit into them. The system was limited to recording small groups of goods. It is also hard to imagine the creation of constant collections, as a handful of small objects is easy to load and it is not possible to keep them in order for a long time. In other words, it was an unsuitable system for use with more complex accounting operations. Clay slates, that is written symbols, became the answer.

The appearance of clay tokens has allowed us to observe three stages of development, which led to shaping the basics of mathematics. Let us return to these stages[54]:
1. Stage one: clay tokens
   – connecting specific shapes with amounts of agricultural goods.
   – the amount expressed in a one–to–one correspondence (specific counting).
   – tokens still did not represent numbers.
   – tokens were not symbols but rather enactive signs. They were one of the new forms of information that had their source in an action–perception relationship.
   – each token accounted for a specific object category only.
2. Stage two: envelopes
   – envelopes were oval token containers with tokens imprinted on their surface to mark what their content was.
   – in this fashion three–dimensional tokens led to two–dimensional signs.
   – the signs on the envelopes were imprinted according to a scheme (a hierarchy–linear format and order was established).

---

[54] Malafouris 2010, 35–42.

- a symbolic representation, made possible by indexing (the establishment of the number line referring to the order of registered objects)[55]
3. Stage four: pictographic tablets
- representation of tokens were inscribed on tablets (iconicity).
- such signs as e.g. "a jug of oil" was preceded by a numeral sign for the first time (separate signs for 1, 10 and 60 were created).
- pictographic signs show the moment when the concept of number was separated (abstracted) from the context.

The path of the development of recording systems in the Near Eastern farming communities shows the serious influence of material culture in creating the concept of number in human cultures. Its story underlines the process of going from material counters to immaterial symbols imprinted in clay, from manipulating material objects to manipulating symbols and finally from a specific system of counting to an abstract concept of number. Moreover, it appears that the system of clay tokens was somewhat like an outer scaffolding, yet a dynamic and constructive one for creating mathematical competence. As a consequence, the protomathematical concepts of tokens were internalized between the human body and elements of reality which were the subject of records. Tokens, which were not yet numbers *sensu stricto,* prepared human minds for the introduction of the concept of number in a society. As Schmandt–Besserat noticed, tokens were a visualization and manipulation of numerical concepts. **In this way the process of counting became full of social meaning and the social relations were filled with mathematical rules**. Elements of the material world which took part in social relations were required for humans to learn its rules. The counting process involving tokens may be seen as a specific interface between genetically based prearithmetic abilities, coded in the human mind and inherited from our animal ancestors, and the material world. Connecting these two elements enabled the understanding of the laws of mathematics.

The use of tokens shaped and enhanced certain neural connections in the human brain. Lambros Malafouris supports this claim with research in which he observed the interaction between parts of the human brain responsible for manipulating three–dimensional objects by the use of hands and those activated when performing arithmetic calculations.

---

[55] The indexicality of meaning indicates that all symbolic messages are dependant from a certain concrete meaning context in which they have been uttered and that is why it is a mistake to analyze their content without it.

According to him, the functional and anatomic relation of these two parts may also be connected to the important role of counting on fingers in the early stage of ontogenetic development. However, these observations suggest that only hand manipulations (fingers) could not have been phylogenetically sufficient for man to acquire mathematical abilities. Moreover, the above mentioned research also shows some limitations of the role of language to understand the first and most basic mathematical rules[56].

In other words, Neolithic tokens did not have to have names to function properly and be understood; they could have been tipped, used and stored unwittingly. As Malafouris writes, objects may represent themselves, thus becoming a sort of missing link between the flexibility of the brain and the flexibility of culture. Traditional methods of description were not sufficient for the growing number of gathered objects along with increasingly complex social relations. And that is why, according to Malafouris, Neolithic tokens could have driven the human mind beyond approximation[57].

The ability to metaphorize, which I mentioned in the first chapter, was an essential element in the process described above. It is not difficult to guess that tokens constituted a specific extension of the human body in metaphor. The protomathematical concepts found their vent in the digit metaphor among almost all cultures in the early stages of perceiving numbers. Expressing numbers was performed through touching and assigning an appropriate number of digits – parts of the body (most often fingers and toes). In many languages which had only a few words representing numbers, a distinct preference for counting on fingers rather than using words, as well as the ability to operate greater numbers than the lexical dictionary would permit, could be noted[58]. It may be said that the system of tokens was the first stage of getting rid of the bodily aspect and expanding the digit metaphor, which was the source of the beginnings of the basis of mathematics. Tokens did not actually represent the objects to which they corresponded on a one–to–one basis as such. They were a material link between the body and the counted object, which earlier was connected to a specific part of the human body. It is beyond doubt that tokens are an illustration of the object collection metaphor, which I described in chapter 1. The context of the counted object disappeared when tokens transformed into symbols imprinted in clay appearing along

---

[56] Malauforis 2010.
[57] Ibidem, 41.
[58] Justenson 2010, 45.

with other goods, and the bodily aspect crumbled with it, thus opening the way to full number abstraction.

CHAPTER THREE

MATHEMATIZATION OF SPACE

By 8,000 BC the first farming settlements had been established in Greece. They consisted of houses, located in a rectangular pattern built of sun–dried bricks, which was a direct adaptation from the Near East. In the later period such constructions were replaced by light wickerwork walls (wattle) and covered in daub. These became characteristic for the whole Balkan Neolithic[1]. About 1,000 years later the first farming societies in the North Temperate Zone started to develop, which were named "the painted culture complex" in archaeological literature. Afterwards archaeologists divided them into archaeological cultures such as Starčevo, Körös, Criş and Karanovo. These cultural societies covered the areas of today's Bulgaria, Serbia, Romania, Macedonia, Albania and Hungary and still contained many common features with the original Neolithic centre from the Near East, displayed in the art factual material[2]. Among others, a similar pottery manufacturing technology was identified there, clay stamps were discovered in many sites as well as the so called *pintaderas* and tokens mentioned in the previous chapter, although their function in the north is still unclear.

The next cultural transformation took place in the north Balkan reaches of the early Neolithic societies based on the development of a new type of existence, whose characteristic relics are vessels shaped as cropped gourds ornamented with engraved bands. The beginning of this transformation can be dated to the 6th millennium BC and it spread to large areas north of the Alps. In archaeological literature these societies have been labelled the Danubian cultures, as the source of the cultural transformations connected with them is believed to have taken place in the Danube basin. Most importantly, the layout of a house was completely different there from the southern design, which could have been connected by adjustment to the increasingly harsh climate. These were long, massive post and beam construction households, which seemingly had no interior division (lack of

---

[1] Whittle 1996, 13–26.
[2] Renfrew 1990, 155.

partition walls). They were built from large tree trunks most often set in 5 rows, which formed a sort of a nave. They could reach up to 50 meters long and 7 meters wide[3].

The colonization connected with the spread of the first farmers above the Alpine belt was quite selective (some speak of an isle colonization), as settlers were mostly interested in loess soils. It was only at the end of the 5th millennium BC that another important cultural transformation took place, which focused on going beyond the loess soil areas and developing different types of farming societies which spread to the furthest regions of Northern Europe, claiming them completely for farming. These have been called the beaker culture or cultures which showed a significant formal similarity[4].

In this chapter we shall focus on settlements and houses, as their connection to the counting and measuring culture seems to be more and more obvious, although still quite mysterious. In the following pages I would like to try to shed some light on this mystery.

Constructions and structures

Settlements and their organization

It is beyond doubt that one of the most distinctive features of sedentary communities was investment in architecture and constructions of different types, built either from wood, which was abundant in central Europe, or stone and wattle and daub constructions, as in the Near East. After many observations in this field we can clearly state that their purpose was more than just the provision of shelter. Archaeologists have put a lot of effort into studying this matter and reached the conclusion that both Neolithic settlements and houses could not be treated as mere shelters, their size and durability being the only factors which distinguished them from Palaeolithic mobile tents. Houses are not only places to live, and a purely functional and pragmatic way of dealing with the climate, but also symbolic structures and even ideological expressions of social desires and fantasies[5]. Others think that houses and settlements should be treated as media of social control and organization[6]. Moreover, all Neolithic constructions may be analyzed on the same epistemological level. Alisdair Whittle proposed that both the Neolithic houses in South Eastern Europe

---

[3] Last 1996, 27–40.
[4] Winiger 1985, 280–281.
[5] Hodder 1990.
[6] Chapman 1990, 49–98; Bailey 1990, 19–48.

as well as the monumental constructions in Northern Europe should be viewed from the same perspective. In both of these cases we are faced with a declaration or demand addressed towards land, which is part of the process of creating a new social identity pattern and a new way of perceiving time and place[7]. Bearing that in mind, let us concentrate on these constructions.

In the Near East we meet two basic systems of Neolithic settlement organization. The first consisted of groups of houses connected to each other, similar to a bee hive; a structure which constantly grows spatially, the famous Çatal Hűyűk on the Konya plateau in modern Turkey being the most significant example of such a structure. Ian Hodder, a long–time researcher of this site, compares such a spatial organizational scheme to that of the American pueblos[8]. The second type was a settlement with spread out houses. Special constructions for communal and religious purposes may be found in some of these settlements[9]. However, it was the agglomerations which became the characteristic type of constructions in most of the Near Eastern area, and in time grew to become the first cities of Mesopotamia. The matter of the organization of the Near Eastern settlements shall be returned to in the final chapter, while below I would like to focus on European settlements.

Early Neolithic settlements in Greece and the Balkans connected with the painted pottery complex were small and had an open appearance. Square or slightly trapezoid houses were built in an area of less than half a hectare, built from bricks that had been dried in the sun and then from wickerwork which was subsequently covered in daub. However, it was at this stage of the early Neolithic period that a visible organization had been introduced. For example, rows of small houses placed in lines facing the same direction and surrounded by an oval paling fence were discovered in the Karanovo I [10] settlement, one of the best examined Neolithic settlements of the region. In the later stage it took on a more agglomerative model, houses had been clustered, and perhaps fences were erected between single households, which changed the original concept of a single direction. However, as Bailey summarizes this early construction period present in the Balkan area, among most of the settlements we can see an organized set–up of constructions, the use of common orientation as well as a distant similarity in size and layout of the constructions. According to him this image provides a climate of a representation of

---

[7] Whittle 1996, 16.
[8] Hodder 2006.
[9] Gates 2003; Wengrow 1998, 783–795.
[10] Tringham 1971; Milisauskas 2002, 182.

common values and with their structure, these settlements reflect strong community cohesion[11].

A Neolithic house and settlement in the Balkan area is, in fact, interpreted as an important social institution which is the theoretical model for the Levi–Strauss' concept: *la masion* – "house communities" or "domesticated communities". The house is a common integration medium for establishing a sort of Neolithic way of life in the whole region[12]. In other words, houses and settlements in the Balkans played the role of active physical structures for social interaction, forming social divisions and maintaining the integrity of the group as well as a means of expression for the whole structure. It is even probable that we should take into consideration the establishment of a new ideology of an anthropogenically constructed environment. This ideology is particularly clear for Bailey because of the insistent repetition of the house scheme as well as keeping to the single orientation of houses in settlements.

The wattle and daub house models–the so called tectomorphs are characteristic for settlements of this period. Apart from these, models of specific equipment, e.g. stove models, which most often appeared in the 5th millennium BC, have been discovered. However, the archaeologists mentioned do not interpret these objects as toys, although they could be considered as such in some circumstances. Bailey believes that these objects suggest that we should perceive settlements and houses, designed in the ideology of the constructed environment of the Neolithic and Chalcolithic periods, at a higher level[13].

In time, some settlements were enclosed with paling fences and ramparts. Houses built in a square or rectangular pattern were set in one row or along straight lines and their constant renewal led to the creation of the so called tells. Tell settlements are the characteristic background feature of the end of the Neolithic/Chalcolithic periods in Northern Greece, Bulgaria, Romania, Serbia and Hungary. These places were inhabited for a very long time (hundreds of years); hence several meters of strata, which contain many archaeological layers, had to be excavated. Tells are a result of the adapted building technique of covering wickerwork walls with a large amount of daub. After the collapse of such a house, its remains formed a thick layer which was not removed, and on it new houses were built. Some settlements have dozens of settlement past horizons. It has been confirmed that this process of the constant renewal of a settlement or its fragments was performed repeatedly after a few

---

[11] Bailey 2000, 269.
[12] Boric 2008, 121.
[13] Bailey 2000, 281.

decades [14]. Tells are an essential feature of the housing architecture manipulation process in the Balkans, which like an optic lens focused on procedures aimed at creating a new environment of human organization. They blend both the "agglomerative" approach as well as the unique innovational solutions concerning common space organization. One of the best examined settlements, as far as this process is concerned, is Ovcharovo[15]. It is located in North Eastern Bulgaria and was inhabited for about 600 years. During the excavation research 13 layers of settlement were identified, with the strata reaching 4.5 meters in height. After a few decades the houses were rebuilt, sometimes slightly changing the inner structures of rooms, sometimes repeating exactly the previous layout of the house. A settlement was always comprised of a few to several households. More than once the settlement or part of it was engulfed in flames or destroyed by floods, as it was built on a flood plain. Hence, the entire layout of the settlement was quite dynamic; although we should point out here that it always oriented according to the same cardinal directions. The basis of this layout was the crossing of two streets lying exactly on a north–south–east–west axis.

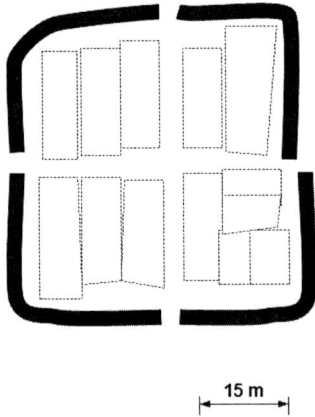

Fig. 3.1 A schematic plan of the Poljanica settlement.

The same was made in other tells of that period. In the Poljanica settlement four entrances leading to it were discovered, while the

---

[14] Ibidem.
[15] Todorova 1978; Bailey 1990.

settlement itself was surrounded by a small paling fence raised in a square, each of its entrances being assigned to a cardinal direction (Fig. 3.1). The Targovište settlement is also an interesting case. It was built on a rectangular plan and was supplied with two opposite entrances/exits, which were placed in a straight line, although in its later stages one of the entrances was abandoned. As a whole this settlement is ideally planned into the crossing of lines which divided it into four parts. The structure of these settlements was apparently a realization of the imagined geometric matrix where its base was the crossing of two straight lines.

Let us return to the tectomorphs for a moment, as some interesting speculations are connected with them. When observing their presence in specific stages of the Poljanica settlement, Bailey noticed their correspondence with the settlement's dynamically changing architecture. It appeared that the number of tectomorphs corresponded with the number of changes and transformations made in the households' structure in the earlier time horizons, which led Bailey to admit that the production of these miniature houses had a strict connection with how the design of the whole settlement changed. This shed new light on the interpretation of settlement structures and houses from the Chalkolithic period in Bulgaria[16].

Bailey formed the hypothesis that the houses of that period were not only places of shelter, feasting and resting but they were also "living beings" as they had their own history, from their birth, growth, life to their death. After the death of each house its clay model, which manifested the past, was created. Houses were remembered and reminisced by molding their miniature models, thus being symbolic continuations of the old households[17].

In addition, the discovery of house models allow us to reflect upon the ways of conceptualizing time in the Neolithic/Chalkolithic Balkan societies. For example we may assume that time was perceived both as a continuum as well as a sequence of events. The new never appeared without a connection with the old–observation made also by Whorf while studying the Hopi Indians[18]. However, it seems that the house models emphasized both the continuum of time as well as the continuum of the society as the organization of space, localization, household equipment

---

[16] However, it is important to mention that house models have also been found outside Bulgaria in settlements other than tellas, e.g.: Lenneis et al. 1995, 91.
[17] Bailey 1990, 44.
[18] Whorf 1956; 2002.

(tools, furniture, objects of everyday use which were inside of them) were directly connected with the social structure[19].

The above information allows us to extend the thesis that, although the Balkan settlements were welfare protection systems[20], they also embodied something more. The emphasis put on the perpendicular setting of main communication roads did not have a direct functional use. This should rather be connected to the concept of social life that was established at that time. It was a period in European history when more complex social structures, which had risen above the egalitarianism characteristic of the hunter–gatherer communities, were introduced. Houses were similar, but not identical. Some of them were distinctively more complex and better supplied than others, which presents us with a better picture of the inner diversity of the society[21]. Nevertheless, the settlement structure reflects a compliance to a universal idea. It seems that the pressure put on living in groups and following social norms, probably introduced through ceremonial activity, was very strong at that time.

The settlement structure had as much similarity with societies as with its imagination of the cosmic order. When planning settlements, a lot of effort had been put into orientating them according to the cardinal directions, or other astronomic phenomena of which we do not know much about at present. Taking this into consideration, we may assume that settlements became more and more a material representation of a new type of society, whose imagination and myths concerning the cosmic world found their reflections on Earth. We shall never know what these imaginations were exactly, but a glance at some ethnological examples would help us to visualize what they might have been.

Cosmological references in household architecture are a characteristic feature of simple societies known from ethnological observations. According to researchers, this was probably the case in the past. In their newest book, Lewis–Williams & Pearce mention the analogies which occur among South American societies. For example the Bororo people from South America perceive the settlement as a structure which organizes the relations of all living beings and natural processes. The settlement's structure is not only based in its everyday life functionality but it also reflects the universal laws which the whole world is subject to. In other words, the Bororo people perceive their settlement as a manifestation of a transcendental reality and treat it as if the cosmos was brought to Earth[22].

---

[19] Chapman 1990.
[20] Ivanova 2008.
[21] Chapman 1990.
[22] Lewis–Williams & Pearce 2009, 93–94.

This is typical for Amazon tribes[23]. Such an approach towards settlement structures was also present among other cultures known from history. Pre-Columbian cities best illustrate this approach as they were built according to the specific directions given by priests–specialists in interpreting the sky and managing the beliefs of the human masses[24].

It is hard to tell how much the imagination of the Bororo people present the essence of the matter. Certainly we cannot use them on a one to one scale. The point is to show that among simple social arrangements, which we believe existed throughout the major part of prehistory, the correspondence between the transcendental world and the real one (material world) was that of the most common mental experiences. This correspondence remained a constant element of later beliefs. Analogies between the human microcosm (the human body, architecture) and the transcendental macrocosm appear in almost all religions. The modern Catholic Church is a local emanation of the community of the faithful as the body of Christ[25]. It seems a legitimate assumption that the deeper we delve into the past the stronger and more complex this correspondence was. Should we not assume that this rule is also present in the form of spatial organization in Balkan settlements? To summarize this matter we can distinguish some characteristic features.

The settlements were built according to a planned construction scheme on a square or rectangular plan. They were surrounded by paling fences or ramparts which could obviously play a defensive role, as Ivanova believes [26]. However, they were not massive fortifications or entrenchments, but were rather supposed to emphasize the social unity of the settlement as a vision of a new type of society[27]. The households themselves were packed tightly and symmetrically and the streets were laid perpendicularly, and it is common to find two or four entrances/exits from the settlement facing four cardinal directions.

The settlement was also a matrix for all other actions connected with its functioning. When looking at the process of constant renewal it seems

---

[23] Zerries & Schuster 1974.
[24] Aveni 1982.
[25] Compare Epistle to the Ephesians 1:22–23: *"And has put all things under his feet, and gave him to be the head over all things to the church, Which is his body, the fullness of him that fills all in all"*. St. Paul in the Epistle to the Corinthians called the community a "house", himself a builder" and Christ the "foundation". *"For no one can lay any other foundation than that which has been laid, which is Jesus Christ."*
[26] Ivanova 2008.
[27] Bradley 1998, 80–82.

probable that a single stage was another act of building a whole new construction anew or an act of subduing the whole structure, whose basic elements were straight lines. Chapman wonders if, as we are aware of these facts, we should consider the Balkan settlements as characteristic acts of the colonization of space. Perhaps they were also structures which, due to their cosmological connection, were supposed to ensure safety?[28]

The tell settlements in the Balkans survived 1,000 years, after which they were suddenly abandoned. Researchers argue over what were the direct causes of abandoning sedentary settlings of such a complex character for the benefit of more mobile living. The later settlements from the 4th millennium BC did not leave much material evidence. The former rules of construction ceased to be valid as societies probably became more mobile[29]. Halting this development, which bore a proto–civilizational character, was connected with the spread of the steppe societies, which in the second half of the 5th millennium BC claimed the Balkans. The effect of this was a change of human lifestyle, which concentrated on abandoning farming for a pastoral lifestyle and leading to a more mobile way of life[30].

Can the spatial organization of the Balkan settlements tell us something of the specifics of a European cognitive society? Let us first see what the settlements and houses looked like in the north.

Neolithic houses in the north

In the first stages of the Neolithic period north of the Alps a new form of settlement had developed, which for the time to come would define the image of a sedentary way of life. Such a settlement was most often comprised of a few to more than a dozen long post and beam constructed households separated from each other by a substantial distance (Fig. 3.2). This was undoubtedly an innovation, although it did not appear from out of nowhere[31]. Let us first focus on the houses. When compared to the southern model, their massive construction is the first feature which distinguishes them. The northern houses had been constructed on solid posts which stood in rows, five rows placed along the whole length of the house being the standard design. Between the outer posts a thin wickerwork wall was spread which, similarly as in the south, was covered

---

[28] Chapman 1990.
[29] Ivanova 2008, 82.
[30] Ibidem, 143–145.
[31] Startin 1978, 143–159.

in daub. As a result of this process characteristic elongated pits appeared along the walls.

The comparison of the long houses' outlines points to the fact that the houses had a module construction, the norm being three modules, although both shorter and longer houses have been found[32]. It is difficult to say something more about the function of specific modules. The central module most often provided the most space and that is why it is assumed that the residents' daily activities were focused there. Joachim Pechtl assumes that a basic module was always built as the first one, to which in time subsequently two modules were added[33]. The outmost modules were built–up much more densely with posts, between which there was little space left, so freedom of movement was limited. One end of the long house could have been the sleeping room and the other–a storage place for different goods and a pantry. It is beyond doubt that such a construction did not leave their residents much opportunity for planning space. An interesting phenomenon is the rare appearance of a fixed space for a fireplace or furnace in "linear" houses, which would be a basic element in the southern architecture designs. Furnaces and fireplaces often appeared outside the household.

As a result of this undoubtedly unconventional construction, there are many hypotheses concerning the functioning of long houses. Eva Lenneis cannot ignore the impression that the "linear" house was built because of its very solid attic in which one could move safely[34]. Other opinions suggest that the massive post construction was connected to the increasingly harsh climate which the farmers faced in the northern areas of Europe. However, it is worth mentioning at this point that the spread of the first farmers in Europe took place during the Atlantic period, which was much warmer and more gentle than the present one. Some, on the other hand, believe that although the linear house looks like a novelty, it must not be studied without considering the connection to the origins of the Neolithic period. Taking into consideration the slightly trapezoid shape of whole houses, as well as each of their modules and their size, connections to the South European Neolithic constructions are clearer. In that case we would be faced here with the far reaching manipulation of architectural elements that were already known before[35].

Houses in the settlement were always oriented in the same direction, although the positioning of the houses shows regional differences (Fig.

---

[32] Modermann 1986, 382–394; Coudart 1993, 114–135.
[33] Pechtl 2009, 186–201.
[34] Lenneis 2000, 383–388.
[35] Startin 1978.

3.2). In the eastern part of Europe houses were most often oriented along the north–south axis while in the west–the east–west axis[36]. Some arguments have been brought forth that such a positioning was connected with the direction of the wind[37], but this does not seem probable. Modderman brings to our attention to the fact that early farmers' settlements had been created in the middle of a forest, which at that time covered most of Europe[38]. Hence, houses were surely not affected by strong winds, which means that their orientation had to have another meaning, although we must remember that functionality did not have to exclude symbolism. Following this trail, Bradley points out that the persistent holding to a single orientation could bear a meaning related to the mythical origin of the linear ceramics people[39].

It is known that linear houses were built only once and were never renewed. Estimates allow us to assume that they were abandoned in favour of a new house built next to it. The position of the old house could still be seen in the settlement for a long time as the house's outlines rarely crossed each other. Here we can see the basic difference from the Balkans, where new houses were built on old foundations creating a settlement which "grew" upwards. An old house in the Balkans had to be demolished, but the memory of it remained in the form of tectomorphs (see above). Meanwhile, the long "linear" houses reminded the settlement of their existence through their physical form, even long after their abandonment. Building tectomorphs became obsolete, although they sporadically appeared in southern areas of the linear societies' existence[40].

There was a lot of free space between the long houses. This is another visible difference from the south. However, these spaces were not completely empty. They were the object of household activity, which can be described precisely on the basis of recent observations. Jens Lüning called this free space a pen (*Hof*) and presented its model[41]. The southern part of such a pen was most often taken up by a large amount of pottery while in the northern part–the remains of stone tools. It seems that pens were divided with regards to their household functions. Several types of agricultural pits were also present in the pen area. These would function as basements, storages, and pantries, etc. Hence, it seems that the settlement model formed in Central Europe at the beginning of the Neolithic period

---

[36] Last 1996, 29.
[37] Marshall 1981, 101–121.
[38] Modermann 1988, 63–139.
[39] Bradley 2001, 50–56.
[40] Lenneis et al. 1995, 91.
[41] Lüning 1997, 70–75.

was more similar to a cluster of individual pens than an organized southern–type settlement. However, this image may be deceptive.

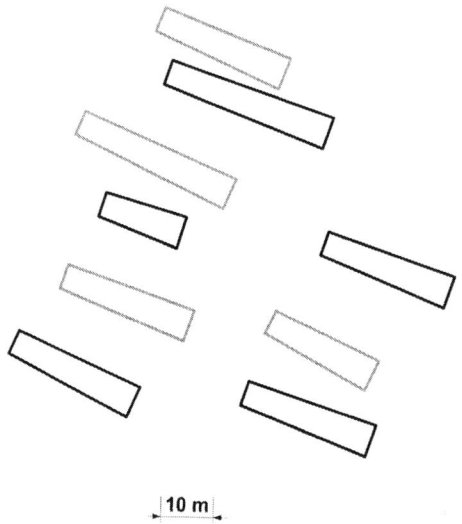

Fig. 3.2 A schematic plan of a Linear Band ceramic settlement.

Symbolism

At the beginning of this chapter I mentioned that Neolithic houses were not only places of shelter, work and rest; they can also be understood as symbolic structures expressing the builder's attitude towards the world. One of their most distinct features in the North was their monumentality. It is a phenomenon which many researchers connect with material evidence of claiming the surrounding soil or a new way of managing space, which was completely different from the one known among the hunter–gatherer societies. In comparison to the light housing constructions of the Mesolithic hunters and gatherers, Neolithic houses appear as monumental constructions which dominated their surroundings. Significant areas of forests were cleared during their construction. Hence, houses had a semiotic value and were a manifestation of new needs, demands and a new social organization and had to be seen as such by the local hunting communities. They were a manifestation[42] of a new attitude towards place, time and nature in general.

The Neolithic house, as a centre of economical and social life, showed the adoption of a new way of thinking. For the hunter–gatherer societies, it was not the tents which were natural cosmologic–symbolic references but the whole world outside of them along with the topography, animals inhabiting forests, the landscape etc. Moreover, they could not accumulate objects in their humble abodes. In contrast, the Neolithic house became both an enormous warehouse containing valuable objects as well as a material representation of the idea of a settled life and perhaps the most significant symbol until the invention of writing[43]. At the same time, the house created a new living environment and grounds for cultivating social relations. Adapting to living in houses the Neolithic settlements must have interceded in the development of the social structure and in thinking about the structure of the world, filling these two conceptual spheres with new meaning. We cannot forget that the symbolic sphere is connected with the social sphere; hence the hypotheses concerning the prestigious meaning of exceptionally long houses of the linear society seem justified[44]. The prestige gained by building monumental architecture probably falls upon the family members who inhabited exceptionally long houses (above 33m long), especially upon the head of the family, but the local society also had its part in this.

Whittle states that the monumental houses of the early farmers of central Europe promoted the formalization of behaviour, emphasized repetition, routine and conformity. The houses referred to a rather open social structure and flexible group composition, they eased integration and tightened bonds in the group. Moreover, he thinks that local Mesolithic communities were involved in the establishment and development of these types of constructions, and that the house was a mark of integration of the indigenous tradition with a new Neolithic ideology. In reality, as Whittle observed, the way the linear house was "packed" with wood seems odd to researchers. He suggests that perhaps we should assume that the linear house is a symbolic manifestation of the acculturation of the omnipresent, and basically a wild, forest?[45]

Ian Hodder proposed to perceive all Neolithic houses as part of the *domus* concept, a powerful metaphor of a settled way of living. While basing this on his study of the Near Eastern settlings, he proposed a

---

[42] Manifestation is an overstatement, a monumentality in action.
[43] Wilson 1988, after Renfrew 2007, 144.
[44] Pechtl 2009.
[45] Whittle 1996, 25.

consideration of the following sequence of events: the house was paved with cobblestones, painted, pargeted and divided functionally/ritually. Since burying the dead or their parts (skulls most often) under the house floors was a characteristic element of farming settlements in the Near East, death had been symbolically brought under the roof and was the subject of control and acculturation that is domestication. Wild animals were also "brought" and "controlled" into the home structure via wall paintings and statuettes. As well as animals, wild plants too were brought into the house and transformed into a cultural product. *Domus* became both a metaphoric conceptual as well as a pragmatic *locus* of transforming the wild into the cultural. In the case of Near Eastern settlements few signs of rituals and material production outside of the household unit have been noted. The house was a sort of material "self–control", which was the source of a broader social control. What Hodder probably wishes to say by that is that the urge for social control had its source in the inner need for controlling the individual, antisocial and wild behaviour present in human kind since the beginnings of using language in society. *Domus* supplied a means of understanding the process of control, and a new understanding of the opposition between culture and nature as well as the social and antisocial[46].

The house, its nearest vicinity and the furnace inside it was a sort of a new microcosm. *Domus* was no longer understood only as a house, as the Latin term used by Hodder would suggest, but it contains a whole set of actions connected with the house (preparation of food, rest), to which symbolic contents are connected. Hence, d*omus* should also be understood as a metaphor of economic and social strategies (most importantly as a presentation of specific authority in regards to the surroundings). It is both the idea and practice, a metaphor of cultivation and care. It is an expression of exclusion, control and domination over the wild outside world[47]. Moreover, it is worth mentioning at this point that *domus* is symbolically connected with woman. On the other hand, according to Hodder, the wildness present outside the house or settlement is *agrios*. It is symbolically connected with maleness, hunting, waging wars and with death, although this division proceeded gradually. That which is present outside is also connected with exchange and social hierarchy and opposes the *domus* metaphor. Although farming was performed outside, in the *agrios* area, it is the symbolic taming of nature and wildness–its domestication.

---

[46] Hodder 1990, 44–45.
[47] Ibidem.

Hodder's structuralistic interpretations are undoubtedly inspiring; hence they should have been given more attention. They became a central plane of reference for many archaeologists. For example Bradley reminds us, in these identification frames that architecture has a specific social meaning and as a consequence of living in it, humans learn to act[48]. They learn that by living with other members of society surrounded by constructions laden with meaning. After all, constructions are so durable that they can be present in the narratives and discourses of future generations. Hence, architecture is placed at the centre of social attention. Architecture is an object of constant interpretation, inheriting many old elements and simultaneously gaining new ones.

Let us summarize at this point. One of the most characteristic innovations in the sphere of construction, performed during the passage from the Near Eastern Neolithic period to the European Neolithic period, was emphasizing the geometric nature of a whole settlement; more specifically building it on an imaginative plan of straight lines which finally began to cross themselves at right angles. The linear form of the house was highlighted in the farther northern area. The rooms were situated one after another in a line. The only effective way (meaning one that would allow a person to move from one room to another) to move in a linear house was generally going back and forth in a straight line. While planning the Balkan settlements, emphasis was put on following the straight lines crossing the settlement and limiting the desires of singular households to sprawl. In the north, the linear house, which was a single farmhouse, became the manifestation of this rule. If we were to summarize the transformation mentioned we may say that **the essence of manipulating straight lines was moved to the north.**

Although Lüning's *Hofplatzmodell* developed for linear societies is still the basic and best known in the literature, it does not determine the end of researching this field. Recently, Olivier Rück presented credible arguments which could confirm that some linear settlements structures resemble Balkan models more than it was hitherto assumed. In some growing settlements he observed some tendencies towards a more strict and parallel scheme of building houses. In this model, which Rück called *Zeilensiedlungsmodell*, houses of a similar date were built tightly next to each other, forming a row in such a way that it was impossible to use any free space around them for farming purposes. Only the front of the house remained free[49].

---

[48] Bradley 1998, 42.
[49] Rück 2009, 159–185.

The building activity in the Balkan settlements remains not fully explained. These settlements were not expanded as much but rather organized anew all the time, while the linearity rule became more and more emphasized reaching its climax in the Eneolithic/Chalkolithic settlements, such as Ovcharovo or Targovište in Bulgaria or Polgar–Csoszhalom in the Carpathian basin, of which more shall be said later on. It is not possible to be indifferent towards the phenomenon of the constant manipulation of settlement architecture. It seems to me that functional explanations cannot match the observed path of changes in the Balkan settlements: moving from unorganized building development, such as Karanovo I, to row building in Karanowo 4 or in tells, such as Ovcharovo, Targovište, or Vinica. These transformations were conscious manipulations of images of the world, such as mythological, cosmological contents as well as those concerning social order (Fig. 3.3).

However, the functioning settlement in the north comprised of several or dozens of "living houses" (the so called currently inhabited) between which the remains of "dead houses" were present in different stages of decay, although these dead houses were for a long time somewhat a living element of the settlement's landscape[50]. It sometimes happened that the younger Danubian societies, later by several years, referred exactly to the outlines of the long linear houses, which means that they were still visible after such a period of time[51]. Eventually, according to Jonathan Last, the end of the tradition of building long households was also connected with the end of the symbolic meaning of the Neolithic house in general[52], and the mystical ceremonial sphere connected with it moved to megalithic constructions as well as to enclosures. I shall elaborate on this process now.

Fig. 3.3 The process of manipulating directions and straight lines took on three basic forms. The essence of these manipulations was building a settlement based on the crossing of two straight lines.

---

[50] Bradley 1998, 44.
[51] Czerniak 1998, 23–36.
[52] Last 1996.

Houses and megaliths

The spectrum of megalithic constructions is very wide and encompassing mould–stone mounds, dolmens (tombs covered with a ceiling slab), cromlechs (circular structures), menhirs (single standing elongated boulders) and many others. It is beyond any doubt that megaliths had a religious, symbolic and ritual function and meaning. Zygmunt Krzak understands megalithism as a universalistic style in culture and architecture, being a reflection of the ancestor's cult and the Great Mother religion[53]. Very often megaliths took the form of tombs in which, for a very long time, people were buried; hence the existence of the ancestor cult seems justified. Although megalithism is a phenomenon of Neolithic provenance, many later Eneolithic societies referred to this type of architecture and customs in different ways. Because of this, megalithism in Europe is a cultural phenomenon that lasted for 3,000 years.

European megalithism can be seen as a compilation of two processes. The first one was a long lasting process of expansion, development and finally the disintegration of the first farming societies, whose main activity was soil cultivation and gardening[54]. One of the basic ideas which were present in these societies, as Hodder suggests, was the *domus* idea and this phenomenon originated from the original homeland of farming, the Near East, where the house was a centre of not only the earthly life, but also the spiritual/ritual one. However, the fundamental difference between the area of Neolithic origin and Europe, which became an area colonized by a new style of life developed in another place, should be pointed out. One of the consequences of this fact were wooden houses. They were a result of adapting to the new living conditions in Europe. The second one–contact with local Mesolithic hunter societies, part of which probably took over farming from their southern neighbours gradually integrated with them[55]. Megalithism, according to many archaeologists, should be understood as a part of this process[56].

In connection to the above, one of the sources of the megalithic idea could be seen in the colonization of vast loess areas by the members of the first farming societies, a stable extension of the settling network, the most visible element being the long house. Because the shape of the first megalithic houses resembles those of the long houses, researchers connect

---

[53] Krzak 1994, 11.
[54] Renfrew 1976, 198–220.
[55] Zvelebil 2001, 1–26.
[56] Sherratt 1990, 147–167; Scarre 2007, 243–261.

these two facts into one process. It is in this concept that Bradley sees the connection between long houses and long barrows–the idea of a dead house transformed into the house of the dead, which in time was taken out of the settlement according to the general trend of separating the living from the dead, of which we shall discuss later on. Recently this process has been schematically outlined by Midgley[57]. In the following stage the long tombs changed into classical megalithic tombs, often with a complex construction and open for adding subsequent burial places. The development of collective tombs had also occurred.

The person who inspired the theory concerning the transformation of living houses into dead houses was Gordon Childe, an exceptionally prolific prehistory interpreter, whose concepts are a basis of scientific discourse to this day. He also proposed the term "Neolithic revolution". Andres Sherratt refers to Childe's ideas by stating that European megalithism was a clash of the indigenous hunters' community with an exotic foreign community of farmers, with the former being in the majority. In other words farmers, who were surrounded on all sides by the native hunter–gatherer groups, had to develop symbolic rules manifesting their presence and the right to own land[58].

While the long lasting and organized settlements remained a constant part of the landscape of South–eastern Europe, the construction of megalithic constructions–symbolic *domus*, material manifestations of the farming concept and land ownership–began in the north. Long barrows were built in the first stage, which were closed from the inside and each designed for only one burial. Some of the oldest of these were discovered in Kuyavia and are known as the Kuyavian barrows (Fig. 3.4). Even older forms were found in the area between the Paris Basin and Brittany in the Seine and the Yonne basin. They are dated to have been built in 4,700–4,600 BC. In literature they have been named long lodges or barrows (*Langbetten*) and they are part of the so called Cerny Culture. 76 types of such necropolises have been identified so far. Their length is their most astonishing feature, sometimes reaching over 300 meters, although these are exceptional. As in the case of the Kuyavian tombs, only one deceased was laid in a straight position along the tomb axis. The subsequent deceased were laid in the built modules of the tomb in one monumental line. Until recently a maximum of four deceased placed in a single monumental line had been noted. Taking this fact into consideration we

---

[57] Midgley 2005, 131.
[58] Sherratt 1990.

may try to formulate the internal chronological divisions of these impressive tombs[59].

At this point we must pause and focus on the feature of linearity and length which was specific to the early constructions. Ian Hodder established the "dramatic idea of linear monumentality", according to which the houses of early farmers were connected to the first megalithic tombs. According to him it was a "central rule for social dominance"[60]. Hodder links linearity with the Danubian tradition, but if we look back at the sedentary structures in Southern Europe, which were characterized by a growing division of settlements through manipulating straight lines, we can extend this tradition a bit further. It seems possible that the idea, or metaphor, of linearity was the subject of social discourse earlier (before the LPC transformation) and it was present during the neolithisation process of the Balkan area. Bearing that in mind, it seems even more probable that Balkan houses were measured similarly to European houses, which we will return to later.

Fig. 3.4 A schematic plan of a long barrow cemetery in Wietrzychowice (Poland).

Ulrich Vait notices that monumental linearity is one of the basic features which characterized early megalithic constructions. He imagines an observer from the past standing before a 100 or 200 meter long monumental mound of earth, for whom such a sight might as well have been a symbol of eternity[61]. Moreover, we cannot forget that the extension of the monumental structures was a process, a cultural discourse, during which permanent forms were constantly developed anew. Another characteristic feature was that their length systematically grew in some regions[62]. Tilley suggests that each form had its own special meaning and it was based on the local landscape, without the knowledge of which they

---

[59] Midgley 2005, 88–90.
[60] Hodder 1990, 233.
[61] Vait 1999, 413.
[62] Voss 1982, 37.

cannot be understood. Local cosmologies developed and were subject to change, the clearest for modern researchers being the long term changes which progressed from the Neolithic period until the Bronze Age, when finally a landmark metaphorization different from the Neolithic one developed[63].

Moreover, Scarre and Sherrat try to complete the southern concept of linearity with the Atlantic element which, according to them, is of Mesolithic origin or at least is not a passive adaptation of the southern model. Its characteristic feature is the regular reference to the circular form, which is already present in the earliest megaliths in Western Europe, especially in the so called passage graves[64]. Scarre specifically underlines the form of the chambered tombs, which is a completely new phenomenon in Northern France where the early presence of the Neolithic economy had a limited character[65]. According to him, the essence of this development is formed by such monuments as Stonehenge.

What connects megaliths with Neolithic houses? In this case, as Lewis–Williams and Pearce present, the amount of the symbolic content reaches much deeper. They suggest that when discussing the problem of the first constructions we should first of all understand the Palaeolithic hunter–gatherer mentality.

Lewis–Williams and Pearce pointed to the fact that the centre of social life for Palaeolithic people was mostly in caves. Not only did they supply shelter but they were also an important element of representation and world view. The paintings of mythological and narrative contents were made on cave walls, which were used as a surface for ritual activities and as a visible element of the cosmos. Whereas the turning point came in the Neolithic period when men began building constructions. These constructions, both tombs and houses, became a materialization of cosmic beliefs. However, what is more interesting, it was possible to control them and, in a way, build personal visions of the universe. This dynamic and active role of man in constructing symbols, images and myths was, in contrast to the adaptive character of the Palaeolithic period, probably the most distinctive trait of the settled human mind. Constructions, whether houses or temples, became the basic means of social control and the people who built them gained great power[66].

As a summary we may say that megalithism was an organizational function for farming production which was a part of the constant conquest

---

[63] Tilley 1999, 237.
[64] Scarre 1992, 121–151; Sherratt 1997.
[65] Scarre 1992, 151; 2007, 257.
[66] Lewis–Williams & Pearce 2009, 167.

of croplands, its legitimization, the development of social control over a group, mixed into rituals, ceremonies and probably trade. The catalyst of the transformation which led to changing long houses into tombs was the relationships with Mesolithic societies. Megalithic graves were often built in groups near settlements, or often in abandoned settlements. Such a localization of ancestors most importantly served as the symbolization of land ownership. The bodies of the deceased present in the tombs, could have been easily accessed by the living, ceremonies were organized near them legitimizing the ownership rights of the social group. When studying clusters of such megalithic tombs, small cemeteries and their material connections with settlements, researchers use them to determine the sizes of local territories. Megaliths were social territory markers which gained a completely new meaning of group demarcation in the Neolithic period, in a dimension unknown before to Mesolithic societies[67]. Their long form was connected with the first stage of marking territories by farming societies while later constructions were the answer from local societies, which accepted farming as a foreign lifestyle. Their symbolic meaning cannot be overrated. Megaliths were a metaphor of a new lifestyle in which the manipulation of architectural form had a basic meaning within this framework.

Enclosures

The monumental constructions of the early farming communities were not limited to long houses and megalithic tombs. Some of the settlements established by those farmers in Northern Europe were in time supplied with entrenchments, trenches or stockades formed in ovals or rectangles. Many of the later settlements were given some sort of paling fence. Bradley thinks these structures are a symbolic expression of society and serve as places of gatherings, celebrations and ceremonies[68]. It seems that earlier establishments were of a rectangular shape, while later on round–shaped structures became more popular.

The first constructions were rarely of a defensive character. It has been observed that they sometimes closed the empty space or cut through existing settlements, as if someone wanted to establish a border between the old and new settlement. Later constructions, which could provide a barrier from attackers, had basically been built only in the later Eneolithic

---

[67] Renfrew 1976, 160.
[68] Bradley 1998, 73–82.

period, while in the first stages only trenches were made[69]. At the Vahingen site we can see that the trench had been built on houses from the earlier stage, hence separating the two parts of the settlement from each other and thereafter used as a burial place[70]. A slightly different situation has been noted in Langweiler 8, where two groups of houses were separated from each other by an empty space which was probably reserved for ceremonial–communal purposes. Later, when people moved to another place the empty space was surrounded with a trench. When analyzing this phenomenon, Bradley notices that these constructions often had a connection with the later stages of the settlements. They were built in places of old clusters of houses[71]. More than once a paling fence had been simply made in close proximity. Sometimes it is difficult to determine its relationship to the settlement.

However, it seems that the first paling fences could have been integrated with the settlements or repeated its plan. In the later period they had been built at a certain distance where they became places of intensified ritual activity, sometimes near megalithic tombs, Calden being a perfect example of this (Fig. 3.5). Calden had been created at a time when settlements generally became less monumental and were comprised of singular small households of a light construction, which were increasingly harder to notice during excavations. This enormous mould–wood establishment has an oval form in which 7 entrances/exits had been built, one of which was probably the main entrance. The layout of the gates seems to follow the geometric rules known at that time. Only the main entrance deviates from this schemata, which probably has a special meaning.

Bradley emphasizes the disproportion between building such constructions as Calden and the tendency to make lightly built houses with a loose layout. He declares that the monumental ramparts opposed the "living" settlements. A large part of ritual activity had been moved to such special structures. What is discovered inside them are the remains of all types of activities, they are equipped not only with pottery, stone tools, animal bones, evidence of the preparation of meals, but also with graves and luxurious objects, which were the subjects of desire and far–reaching trade. In the context mentioned above, such establishments are interpreted as places for ceremonies, festive meetings, performing rituals, negotiations, trading different objects, arranging marriages etc., although

---

[69] Trnka 1990, 213–230.
[70] Krause 2001, 109–136.
[71] Bradley 1998, 45.

their defensive function is not ruled out[72]. In other words it seems that the settlement ceased to be an area of ritual activity as such in the Eneolithic period as this role had been taken over by constructions built for this particular purpose. They referred to old traditional settlements, performing an integration role, and for society bonding. Lüning reminds us, on the basis of the special analysis of the Isar and Danube basins, that these constructions were socio–ritual centres connected by a common communication network [73]. Five such constructions, which probably created a common area of communication and trade, have been recorded there. Rondels and enclosures had become very popular in the Eneolithic. An estimated 38 of these have been recorded in a small area of Eastern Austria.[74].

Enclosures were also supplied with entrances facing four cardinal directions or placed in such a way to be inscribed into a network of straight or perpendicular lines (Calden). Sometimes the constructions which stood inside the building corresponded to the whole through their geometry, Polgar–Csoszhalom, which was surrounded by five massive concentric trenches, is an ideal example of this. A few constructions situated radially with their longer axis pointed towards the centre of the settlement. The central construction stood there, inside of which very interesting discoveries were made. I shall write in more detail about them in chapter 5. Researchers who studied the Polgar–Csoszhalom site think that it played a major role in ceremonial exchange between communities living in the Carpathian Basin at that time. This is confirmed by the materials found at the site which come from different cultural groups[75]. Yet again we face factors recently described in this work. Linearity (although entrenchments are circular, their form is based upon a distinctively fundamental network of straight lines crossing each other) was still the basic rule of organizing these structures. The constructions completely follow the general geometric layout.

Here we can see a symbolic structure similar to that of the megaliths. As well as long graves referred to the image of the long house of the early farmers, the ramparts could be a metaphor of the whole society, the whole settlement. When taking into consideration the development in the time mentioned above it may be said that they were a symbolic representation

---

[72] Petrasch 1990, 407–564.
[73] Lüning 2005, 72.
[74] Lenneis et al. 1995, 82.
[75] Raczky et al. 1994, 231–236; 1994.

of the settlement's idea–the mythical settlement [76]. Anchoring it in cosmologic relations via straight lines made it refer to ancient tradition.

As I have already mentioned, the idea of the rotundity/circle is probably an element of an Atlantic world[77]. In Western Europe, when megaliths were being built, it appeared early and clearly which is visible in the example of home architecture such as monuments of the Cerny society in Western France. Studies show that both long houses, referring to the Danubian tradition, as well as round houses, which were most often internally divided in two parts, functioned in these societies. The idea of linearity and rotundity is mixed in an equally vivid way in the sequence of building tombs. In the first stages, long graves of a slightly trapezoid shape had been built. It was a trend which reached its peak with the Passy type which reached up to 300 meters in length! In the final stage, graves had been finished with a round mound in the shape of a burrow or a round paling fence[78]. Seweryn Rzepecki sees an effect in this confrontation of linear tomb ideology, which comes from the Danubian tradition, with the lithic Mediterranean models, which he called the megalithic synthesis[79].

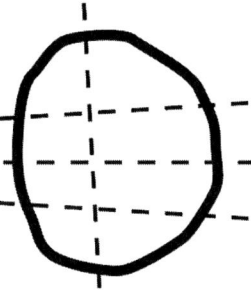

Fig. 3.5 A schematic plan of the Calden enclosure
(according to Raetzel–Fabian 2000, reworked by the author).

Measuring for the living and measuring for the dead

Attempts to understand the megalithic idea resulted in researching methods of construction. Researchers sought the answer of how small technologically primitive farming societies could build such massive

---

[76] Bradley 1998, 82.
[77] Ibidem.
[78] Midgley 2005, 90.
[79] Rzepecki 2011, 220.

constructions. Was this possible without any planning, engineering, mathematical and measuring abilities? Could astronomic observatories, for which some of the megaliths are considered, be built without performing complex measurements?

In 1936 Schröder focused on the proportions of some of the megalithic tombs and using this as a basis he tried to show that during their construction measuring strings had been used and the Pythagorean theorem was known. His ideas were taken up by Helm (1952). He tried to prove that marking straight lines between the outlines of Neolithic house posts (perhaps by the use of a string), as well as the size of houses, reflected simple mathematical proportions[80].

The breakthrough in searching for mathematical regularity in the Neolithic period came in 1955 when Alexander Thom, an engineer and professor at Oxford University, published an article entitled *A statistical examination of megalithic sites in Britain*[81]. For this article he performed measurements for the diameters of 46 megalithic circles, which had appeared in Britain by the end of the 4th millennium BC, and which were built deep into the Bronze Age. Thom recorded the values of these diameters on a graph and came up with some very promising results[82]. On the basis of his results he assumed that during the building of mystical constructions some sort of measurement had been used because the maximum values of the frequency layout for the diameters seemed to concentrate at regular intervals on the graph scale. Thom claimed that some metrologic standard, which he called a megalithic yard, had been used to determine the placing of stone circles. His calculations show that his measurement equalled about 2.72 feet, which is about 82cm. Thom also tried to prove that the megalithic yard was subsequently divided into halves and quarters.

However, Thom's research met with heavy criticism as it was hard for archaeologists to acknowledge that there had been a single measuring standard in the Neolithic period which was used in the whole of Europe. All of the previous sources clearly pointed to the fact that such a standard came into use in civilizations which possessed a developed bureaucracy and government, which could impose certain norms. This case caught the attention of a statistician by the name of Kendall, who used the occasion to present his own method of searching for metrological structures[83]. In 1974

---

[80] Rottländer 1999, 189–202.
[81] Thom 1955, 275–295.
[82] Thom 1967; Renfrew & Bahn 2002, 383; Renfrew & Bahn 1990.
[83] The first method was presented by Broadbent (1955), 45–57; compare Dzbynski 2004.

he criticized Thom's theory saying that only a small percentage of the measurements made by Thom can be counted as statistically relevant[84]. Other critical analyses followed, most often proving that the megalithic yard is quite an elusive phenomenon in the statistical sense [85].

Thom's measurement may indeed cause confusion. Most importantly it is hard to imagine that our Neolithic ancestors needed to measure stone constructions with accuracy to two decimal places as the MY measure suggests. This measurement should be treated with appropriate balance. It surely was not focused around such accuracy. On the other hand, if a person builds something it should not be surprising that the person is able to perform simple measurements. Some simply proposed that for such enterprises as megalithic constructions measuring lengths using steps would suffice. Hence, we do not have to imagine developed measurers, who would walk with a pair of compasses and a standard measuring stick, but the use of a string or a makeshift fragment of a straight stick should not be excluded. Nevertheless, the pace measure is also a measurement, which in certain conditions would be sufficient to plan and build such constructions as megaliths. We also cannot exclude the option that this measurement served not only as a way to measure, but also to manifest the activities connected with the process of measuring. In other words, the focus may not have been put on the necessary precision but on the awareness and knowledge of the concept, symbol and metaphor which connected monumental constructions with the universe. **In short, the purpose of the measurement was not only to be precise but to give meanings**.

Over twenty tears after Thom, an amateur archaeologist Werner Rasch took up the engineering concept and used his method to study the outlines of households belonging to the oldest farmers in Europe[86]. He extended Thom's hypothesis for Neolithic houses stating that also in this case this measurement had been used, inter alia when determining the length, width and even the internal division of houses. Here as well, as Rasch assumed, the basic measurement (Rasch renamed it the "Neolithic length": *Neolithische Länge*) was used and based it on halves and quarters. He then used his methods to study megalithic constructions in Mecklenburgy and Lower Saxony, where he reached identical results.

The outlines of Neolithic households are probably easier to measure than stone megaliths because wooden posts are normally much smaller than massive rocks, but even here we are faced with certain difficulties.

---

[84] Kendall 1974, 231– 66.
[85] Davis 1983, 7–11.
[86] Rasch 1987, 341–346.

That is why Rottländer assumed that the measurements were made on the outer side of the house and for this purpose a stretched string had been used. He brought linear houses' outlines onto a measuring grid, whose basic measuring unit was the "Neolithic length" calculated by Rasch. Although the linear houses differ in length and width, Rottländer discovered that the most often used ratio was 1:4 which means that the length of a house was four times longer than its width. And so the most typical LPC house had from 46 to 50 NL (*Neolithische Länge*) and from 9 to 15 in width[87].

When focusing on the concept of the Neolithic measurement system, Rottländer tried to emphasize its meaning in a broader context. He presented the concept according to which length measures initially had different aspects, although they always originated from perceiving specific body parts as a comparative standard. In time a single standard had developed and it spread, along with the expansion of farming. Rottländer also proposed that the source of European measurement should be sought in the birthplace of farming, which is the Near East. He believed that its direct source are the examples of Near–Eastern measuring, evidence of which remains to this day, called the royal cubit, which had been used as far back as 3,000 BC. He even calculated the exact numerical correlation between cultural centres distant from each other, namely 10 megalithic yards equals exactly 16 cubits[88].

An omnipresent yard?

Let us return to the Balkans where we noted that the characteristic rules of planning settlements, which were based on respecting straight lines, had been followed. Studies which aimed at raising the thesis concerning the Neolithic measuring unit present in Neolithic houses in central Europe inspired Nikolov, a Bulgarian archaeologist, to concentrate on some Neolithic households in Bulgaria[89]. He focused his attention on Neolithic houses at the site in Slatino situated in the suburbs of Sofia. At this site a settlement from the Neolithic period (Karanovo I stage) was discovered, inside of which several houses characteristic of that period were found. An additional prompt was the very good state of the houses, as some of their walls remained up to 70cm high. In such conditions precise measurements could be performed.

---

[87] Rottländer 1999.
[88] Ibidem.
[89] Nikolov 1991, 45–48; 1998, 115–124.

The Slatino houses were rectangular or slightly trapezoidal in shape, the shorter walls–the southern and northern–were 9.69 and 9.27 meters long, the longer walls were almost of the same length–12.44m and 12.34m (Fig. 3.6). The house had two rooms, a smaller and a bigger one, separated by a wall. The smaller one basically was formed in a narrow (about 2m) corridor, but it was placed on the opposite side to the entrance. The walls were built from oak pole mixed with nut–tree branches and then plastered with clay. The roof, probably a gable roof, was supported by inner posts as well as the walls. Inside the house a total of 27 pieces of tableware and construction elements were found: fragments of grain silos, a furnace, a fireplace, a weaver's tool, stone querns and a cultural niche in which a miniature model of the house had been discovered[90].

Trapezoid houses were not an exception in the Balkan Neolithic period and Nikolov assumes that apart from its functional meaning, which is not fully known to us, the shape had to be achieved according to a plan by measuring the wall lengths. Hence, in the case of the house studied in Slatino it should be assumed that those measurements had been made. The northern wall is shorter than the southern one and Nikolov assumed that if any measurements had been made on these walls then the difference in lengths should reflect some multiple of a measurement, which would mean it was not random. Eventually, it appeared that the difference in length between these walls was equal to (9.69–9.27 = 0.42 cm). If this value is accurate then it should also be reflected in the remaining lengths of that house. In the end Nikolov did present the following results: $9.69/0.42 = 23.07$ and $9.27/0.42 = 22.07$. The remaining 0.07 Neolithic measure, which is evident, constitutes only 0.03 of a meter and can be surely ignored. Therefore the measurements of the Slatino house seem to show that it had been planned with the use of a multiple of a length measure = 0.42m. Moreover, if we were to draw a straight line exactly through the middle of the house in Slatino we would note another even value 29, which would be equal to 12.18 of a present day meter. Even the corridor placed at the end of the house is measured to be 3 units of the Neolithic measurement ($3 \times 0.42 = 1.26$m).

Those 42cm are almost equal to half of a megalithic yard proposed by Thom, but I do not assume that combining both of these results is required. Despite this, Rasch, after studying several Neolithic and Eneolithic sites in Bulgaria, reached the conclusion that the measurement used there was equal to 83cm[91]. As I have already mentioned, I do not aim to show the

---

[90] Ibidem.
[91] Rasch 2000, 33–46.

unification of measurements in prehistoric Europe in this thesis, but that constructions were measured in a specific way by the use of a linear measurement, which probably had some reference to the human body is sufficient.

Another example comes from Slovakia. Karlovsky and Pavuk analyzed several Neolithic houses and the enclosure–shaped fortified mould construction, concluding that during their construction a length measure had also been used, which they called the lengyel fathom. When performing their analysis they also came to the conclusion that this fathom was also a relatively fluid measuring standard as its length ranged from about 1.87 to 2.07 of a present day meter. In the houses' outlines, as well as in those of the pallid fences which accompanied them, they noticed a doubling and halving of that length measurement[92]. How can we explain such visible inaccuracies?

Slovakian researchers rightly assumed that early length measurements had a strong connection with parts of the human body, which means they were anthropogenic measurements. In the case of the lengyel fathom, its most probable reference would seem to be the length of the human body plus an outstretched arm. If we assume that the average height of a human body in the prehistoric period was much smaller than today and was estimated to be equal to 157–160cm, a man of that size with his arm stretched upwards would in fact match a fathoms length of 194–202cm. Differences in the fathom length had been noticed in Slovakian sites, which means they must come from people who differed in height.

The Slovakian archaeologists' studies provide us with justified grounds to perceive the early length measurements in the European Neolithic and Eneolithic periods in such a way. We may imagine that each construction planning, whether of a monumental house, a long grave or other type of megalithic construction, was preceded by "taking the measurement" of a certain human being. It was undoubtedly a ritual event, just like the whole process of making measurements and the construction itself, which we cannot imagine without the cooperation of a part of society. The process of measuring and building in the Neolithic period probably had a social and religious character.

It would suffice if, after acknowledging the above mentioned pieces of information, we accept the thesis according to which simple measuring methods were actually used when building constructions in Neolithic Europe. We do not have to argue about the precision required to perform such a task nor if the same standard was used everywhere. If we accept

---

[92] Karlovsky & Pavuk 2002, 137–156.

this assumption, which is basically logical, we may also accept the following: in all of these instances we are dealing with a manipulation of a straight segment which, in a parallel way, was a numerical representation. These actions were connected with manipulating **the concept of the measuring stick**, whose cognitive meaning was presented in chapter 1. Before we proceed to the concluding interpretations let us discuss the matter of the correspondence of megalithic constructions with astronomic phenomena.

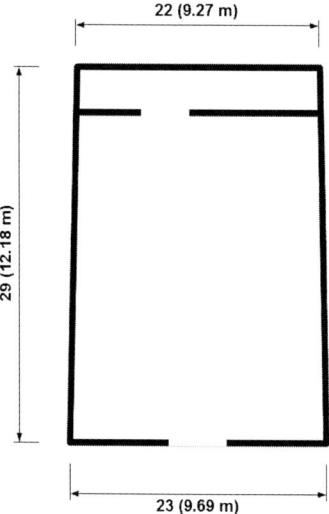

Fig. 3.6 Measurements of a Neolithic house in Slatino (according to Nikolov 1991, reworked by the author).

Archeoastronomy–(measuring–don't leave home without it!)

The credit of archaeologists who researched the post–processual tradition paid attention to the relationship between society and the landscape[93]. Astronomic events were also part of that landscape. Let us briefly focus on this matter.

Systematic research performed over the years in reference to the megalithic constructions present in Western Europe, from France to the shores of the Mediterranean Sea, allowed some systematization of this matter. Over 3,000 objects have been studied. It appeared that none of the monuments pointed toward the northern quadrant, in which (over the

---

[93] Tilley 1994.

horizon) none of the main celestial bodies (the Sun, the Moon and the visible planets) ever appear. The orientation of the monuments most often corresponded with the direction of the rising Sun and its movement to the zenith. The direction of the descending Sun was very rarely chosen. The results of this research clearly point to the fact that the Sun and its movement was the main indicator for constructing megaliths. Researchers also call upon works in which attempts were made to orient lunar directions connected with the Moon's movement. However, these were very difficult to analyze as the movement of the Moon has a very complicated trajectory[94].

According to the majority of researchers, directions were mostly connected with the rising and descending Sun and its position in the sky at certain days and seasons. There are four such crucial direction–dates: two solstices and two equinoxes. The most popular, and at the same time one of the most massive megalithic constructions which confirms the astronomic construction rules, is Newgrange in Ireland. During the winter solstice (midwinter) the beams of the rising Sun fall through a specially placed window above the entrance to a 19 meter stone corridor and remains there for several minutes. During that time it covers the only ornamented stone placed at the end of the corridor with its light, on which spirals, zigzags, points and straight lines had been placed. The window above the entrance could be shut and opened. It is beyond doubt that this was planned[95].

The most popular megalithic construction is, of course, Stonehenge. It has undoubtedly much in common with prehistoric astronomy. It is a very complicated construction which is comprised of ramparts, trenches, stone circles and evenly spread holes called Aubrey holes. These were made with great accuracy, creating a circle 86.6m in diameter made assumingly with small posts and a measuring string. Inside there was a horseshoe–shaped stone construction made up of five massive stones–trilithons, which are structures composed of three stones, two of which stood beside each other and the third lay horizontally on their tops. This whole construction resembled a gate with a narrow passage. The weight of each stone is up to 20 tons.

Not all of Stonehenge's elements were erected at the same time. The construction of this sanctuary may be divided into several stages which stretched over 2,000 years. What this means is that many culture groups participated in building this construction, starting with the Neolithic and

---

[94] Ruggles 2012.
[95] Lewis–Williams & Pearce 2009, 231.

finishing in the Bronze Age. It is worth noting that one of the earliest elements of this structure are those precisely measured being the Aubrey holes, inside which human remains have been found, dated to the end of the 4th millennium BC. On the other hand, the second stage is connected with artefacts of the beaker culture which spread at the end of the Eneolithic period throughout nearly all of Europe, probably introducing the custom of drinking alcohol and warrior behaviour to local cultures[96].

What was Stonehenge? Its remarkable construction has already raised a huge amount of assumptions and guesses in the past, each surpassing the other with its ridiculousness. It is in the light of academic knowledge that we may speak of two certain functions: sacral and astronomical. Both were strongly connected to each other. The sacral character of Stonehenge results from its structure. It was built on a circle scheme with a sanctuary in its centre. The circular construction can be treated as a sacred place, known from the oldest of times, which separated the sacrum from the profanum. What also speaks for the cult character of Stonehenge were the burials which took place nearby and as well as fragments of human bones (perhaps sacrifices) in the circle itself. No signs of prehistoric settlements have ever been found close by.

As to Stonehenge's astronomic character there is not much doubt. The fact that the main axis of the construction is aligned towards the point the Sun rises during the Summer solstice had been noted in the 18th century. The remaining stones indicate other important days of the year which are connected with key astronomic phenomena. Many other analyses had been performed, their aim being to determine whether the construction was an accurate astrologic observatory. It is a complex problem because, as Krzak explains, in Stonehenge there are 27,060 directions marked by 165 significant points. On the other hand, there are many celestial bodies in the sky which could have interested our ancestors. Most of these analyses suggested strong solar and lunar correlations which would mean that Stonehenge served to "include" the most important dates in the calendar, whose basis was the observation of the Sun and the Moon. According to Krzak many of the astronomical connections of this stone construction pointed out by researchers suggest that it also served as a calendar with the year divided into 4, 8 and 16 parts[97].

Knowledge and skill were definitely required during the construction of megaliths. Specialists probably existed, which were experts in planning these constructions and putting those plans into action. Simple wooden,

---

[96] Sherratt 1987, 376–402.
[97] Krzak 1994.

horn and bone tools had been used as well as stone axes (later metal), wedges, ropes, etc. Many experiments and observations have been performed which certify that the construction of megaliths took place using the resources available at that time. It has been estimated that eight people were needed to move a 5–8 ton boulder, about 200 people equipped with ropes, about 30 wooden rollers and several poles, used to steer the load, to move a 32 ton one[98]. An alternative source of studying prehistoric construction techniques is also the iconography of ancient civilizations. In Egypt special sleighs and appropriately prepared roads had been used for large boulders. Water transport had been used whenever possible[99].

In the case of more complex tasks, such as stone circles, large corridor tombs etc., simple measuring tools were probably used, as these kinds of constructions could not be erected "by guessing". Hence, many researchers, who focus on the problem of megaliths, acknowledge that Thom was right that a material model of a measuring unit, that is an equivalent of today's meter, had to exist even if it is hard to prove it statistically. However, the question whether the same model had operated in the whole of Europe for thousands of years, is a separate one and we do not have to address it here. It is surely difficult to agree with Thom and his followers that the measurements were made with an accuracy down to centimetres, which is suggested by the value of the megalithic yard, sometimes given with an accuracy up to the third decimal point! It seems that an anthropogenic measure, which may have been set before the construction began, was simply used.

Conclusions

Perhaps the problem of accepting the megalithic yard is connected with the fact that there is a lack of a theoretic and holistic approach which would make it possible to direct this problem more towards the centre of scientific discourse. I would like to outline such a theory in this work. Most importantly, it seems that the megalithic yard concept is treated too literally. In reality it is difficult to acknowledge the assumption that during the Neolithic period an identical 0.82m length measurement was used on the whole continent as this would leave archaeologists with a vision of a highly developed society and its central authority, which was able to impose measurement standards. It requires merely a glance into the not too distant past, when different local weights and measure standards were used

---

[98] Renfrew 1976, 142–146; Krzak 1994; Renfrew & Bahn 2002, 302.
[99] Romer 2007, 187.

even between small areas, to see that trade required a constant recalculation of one measurement to another, rendering the trader into a true arithmetician[100]. Only ancient authoritarian nations (although not always, of which I wrote about in chapter 2 could show a certain homogeneity, but are we allowed to use these types of political solutions in the Neolithic period? If we had to search for such analogies then we could rather expect cultural–regional divisions, such as those observed in other elements of material culture, not the central managing of metrological standards. However, perhaps the assumption concerning the political or social isolation, the cultural breakdown of the Neolithic period, should be revised? Perhaps we do not need to imagine a centrally managed authority, as a certain cultural unity could have been achieved by general communication and exchange of thought, as archaeologist Jens Lüning suggests?[101]

The problem with accepting the idea of prehistoric measuring may be linked to the way it was presented. Perhaps the question of whether the megalithic yard or "Neolithic length" actually existed is inaccurate. In my opinion it is so. At this point I would like to refer to Kripke's arguments, which I mentioned in chapter 1. This is all the more visible as the matter of the measuring stick was also broached by him. Hence, he points out that the difference between stating that the stick has a specific length and that it has a specific length at a given time is practically none. Therefore, the fact that the stick has a given length at the moment does not make the sentence true. In other words, any number on that length will be true as long as we find its use in social communication or paraphrasing Wittgenstein, as long as we include its role in the "measuring game". We must keep in mind that following a rule only has any sense through a social perspective.

Kripke claimed that by using the definition of a measuring stick of a specific length we do not give meaning to the stick's definition but we only find a reference. If a certain length, which we want to determine, exists we may determine it by using a casual asset; namely we find, for example, a stick which matches our desired length. Another person could also determine the same reference by using some other random asset[102]. Of course this does not explain why the measures identified so far are quite similar but it suggests the important role of the human body, which is a source of perceiving reality[103]. After all, bodies were different and so were the reference situations. As a consequence it is impossible to identify a

---

[100] Kula 1986; 2004.
[101] Lüning 2005.
[102] Kripke 2001, 76–79.
[103] Merleau–Ponty 1945; 2001.

Neolithic measure by using precise statistical methods, which are based on the existence of a single permanent unit. There were as many megalithic yards as there were bodies to which they referred. Without taking into consideration the bodies of prehistoric men, these areas of research are bound to fail.

The observations mentioned above and research put Neolithic constructions in a completely different light, although still more questions than answers remain. What results from them is, first of all, that the issue of the existence of the metrological concept in the Neolithic and Eneolithic periods must be put into a social communication perspective[104]. We are not faced here only with measuring but also with finding references to many other matters of which not much has been said before, such as ideology, religion, authority cosmology, etc. Erecting megalithic constructions indeed was a part of those social measuring games at that time and have been disregarded in present day archaeological discussions. The echoes of this approach are still seen in the matter of today's measures, for example the meter model of the time which Wittgenstein and Kripke were interested in. Although they are placed in a far more abstract perspective, the references towards the authority of social order are clear even in this case.

The planned subordination of architecture to number and measuring did not only concern sacral objects but also houses, although we must remember that this division is highly artificial. The separation of these two categories is not justified in the light of the interpretations presented as not only ethnologic analogies but also numerous archaeological materials clearly suggest that Neolithic households were filled with deep mythological and ritual meanings. That is why we may assume that the supposed Neolithic measure and its uses had references to religion and ritual. Cosmological references were at that time present in every construction, which resulted in the need for specialists who planned such constructions. Perhaps they were also the first people who could add, subtract and use mathematic relationships in practice.

Apart from being a measuring game, it was more importantly a measuring ritual which suggests that the measuring stick metaphor was deeply fixed into narrative contexts. As David–Williams and Pearce point out, during the period in which narrative content based in the Palaeolithic still had to be alive, a conversion of cognitive structures had to occur. Early settled societies were the first ones in human history which constructed their symbolic communication systems so dynamically by the

---

[104] Dzbyński 2008.

use of architecture. The basic frame of reference for these symbolization systems were settlement structures, which served as external symbolic storage networks. This architecture became a powerful communication medium which defines the way of living to this day, specifies our social institutions, frames of reference and creates a stage on which we play out our social relations and roles. We must assume that in the Neolithic period, and in fact at the end of the Palaeolithic period, as discoveries in Göbekli Tepe[105] show, people were perfectly aware of the symbolic dimension of all constructions[106]. This theorem refers to Jaque Cauvin's work, who, while contesting the materialistic–processual approach, claimed that the so called Neolithic revolution had appeared most importantly in people's minds long before they became farmers[107]. Caves, which for tens of thousands of years were centres of social and religious activities, had been replaced by anthropogenic constructions which had to be filled with the proper symbolism. Building had become a representation of the cosmos as well as its manipulation.

In the above mentioned context it is very important to be aware that architecture, being the new media in presenting socially important messages, is not truly a neutral medium[108]. Building meant practicing the manipulation of measurements of length which bound earthly life to religious and cosmological concepts. Because of the above it seems justified to accept the hypothesis that the metaphorization of the measuring stick became an important cultural subject at that time, while the abilities and knowledge connected to using measures became a tool for manipulating the universe as well as society. Chapman writes that the steady geometric order observed in the Balkan settlements apparently served as a means to control society by blocking the urge of single houses to overly spread. This type of social control had to lie in the hands of the family elders, who could have had exclusive power to implement these plans into settlement life[109]. However, this order was very much based on the linearity of space rather than on its fragmentation, which is emphasized in the Near Eastern settlements[110]. We should not be deceived by the idea that such simple actions as measuring by use of a piece of wood or a string did not require special abilities at the time they began

---

[105] Schmidt 2006.
[106] Watkins 2004, 5–23; 2006, 15–24.
[107] Cauvin 2000.
[108] McLuhan 1964; 2004.
[109] Chapman 1990, 83.
[110] Gamble 2004, 91–92.

their path of development. Perhaps it is only a paradox that this process is best seen in Europe, of which we shall discuss later.

If we assume that the concept of stages of reaching the mathematical truths, which I presented in chapter 1, is true then people in the Neolithic period were at the beginning of this path. Manipulating mathematical concepts was surely not so easy and abstract as it is today. We should rather assume that it did not yet separate from the concrete, which several virtually different measures discovered in Europe seem to suggest. Their illusive similarities probably result from the fact that they were anthropogenic measures taken on the spot. Although it was possible that people were able to manipulate larger numbers, we may also assume that counting did not take place in the memory of a single specialist but was connected with **working in a group and had a strong connection with the body, perhaps with many bodies or with different body parts which were a reference for different areas of myth and ritual.**

These observations allow us to all the more perceive early mathematical achievements as complex cognitive and social actions. To count and measure people they had to use complicated concrete counting and measuring techniques. Both counting and measuring required object manipulation in that era, as well as manipulating people. Hence, we may imagine building and planning settlements, houses and more importantly cult constructions as communal events during which specific people participated in the process of measuring, counting and simple arithmetical operations, while one or two people managed their actions. **Therefore, measurements reflected in Neolithic households and megalithic constructions emphasized the essence of community.**

The measuring stick metaphor: a segment, line, object, which may be also a number, probably became the first stage in the mathematization of space and social relations in Neolithic Europe. It seems that a different division of accents had occurred in the Near Eastern Neolithic, of which I shall write about in chapter 6. The measuring stick, if it was used during the construction of such settlements as Çatal Hűyűk, first appeared in the background because of another medium–recording systems, which I explained in the previous chapter. While in the European Neolithic and Eneolithic the measuring stick was a significant contribution to new mathematical thinking. The metaphor rooted itself so deep in the Neolithic and Eneolithic communities that it was able to adopt its next form. It influenced the development of tools, becoming more personal, and becoming entwined deep into individual human relations. The monumentality of the first constructions, which connected the living and the dead in one consistent metaphoric message, underwent sublimation and different alterations. One of these were macrolithic tools, which I will demonstrate in the next chapter. Another will be the growing cemeteries separating the living from the dead as well as the whole community from nature. Hodder's interpretation suggests that in this fashion the almighty metaphor of domestication, still so monumentally present in the early Neolithic period, later on became sublimed and defragmented in social relations. However, it was the time for the new technology of metallurgy which heightened the resolving of the first Neolithic metaphors and symbols into new, more individualized and abstract metaphors of the Metal Age.

CHAPTER FOUR

THE MATHEMATIZATION OF HUMAN RELATIONS

The most recognizable example of the mathematization of human relations is the process which went from tokens to the creation of accounting in the Near Eastern civilization, of which I wrote about in chapter 2. However, let us remember that this system did not spread in Europe. Although Michael Budja tried to demonstrate that tokens played a certain role in Balkan Neolithic communication, there is no evidence of the existence of a material form of the protonumber (reference in a one to one correspondence) in social relations. Does this mean that for a very long period of time, which was till the Bronze Age as only then artefacts had become "clearly" connected with counting and measuring (e.g. weight), people in Central and Northern Europe lived and functioned in blissful ignorance of any mathematical concepts?

From the previous chapter we know that it was probably not like that. Mathematical concepts appeared at the start of the Central European Neolithic in the framework of a powerful communication medium, namely monumental architecture constructed with the use of manipulation of the measuring stick metaphor. We must remember that the measuring stick was also a metaphor where meaning covered the concept of number and cosmologic connections as well as those which concerned social order. Abstract numbers were not known at that time, although by the use of the measuring stick some basic mathematical manipulations, such as halving and halving a half, which is quartering, could have been performed. Adding and subtracting was also possible. Mathematical rules also appeared, although they had strong bases in the societies' narrative sphere (in myths, rituals and stories which concerned ancestors and nature) as well as a strong connection with the context of performance. We may assume that the concept of number used when measuring megalithic constructions did not have to be compatible with the ability of counting/measuring other objects.

The perception of number (protonumber) in the Neolithic period was characterized by a certain distance. In architecture, in which language was the measuring stick metaphor, was used to emphasize the connections in a human group, connecting ancestors located in megalithic tombs, cosmologic beliefs and society into a single coherent system. However, it was not able to introduce the concepts of number into direct social relations. They emphasized community rule. This is why the early farming societies appear to us as quite homogeneous, while the rules for expressing social inequities were quite limited. The same houses were constructed and the same tools and vessels were manufactured from the Paris Basin to Volhynia over hundreds of years. The monumental house of the post and beam construction had been placed in the centre of social attention and it was replaced with the megalith in the later period. However, along with the introduction of new concepts from the South, the measuring stick metaphor had been enhanced with a new interpersonal character. Let us take a closer look at this story.

Early exchange systems

When speaking of social relations, we may acquire a lot of information from studying graves. Along with the spread of farming to Central Europe burials had been placed more and more often outside settlements, which is one of the most visible processes of cultural change in the Neolithic period. This is clearly visible in the Balkans, where in some tells a part of the deceased were buried (sometimes in an unoccupied area of the tells), but establishing large cemeteries near settlements was a definite change[1]. The best known of these are Durankulak, Ruse and of course the Varna I cemetery, where more than 280 graves were discovered. As far as northern areas of Europe are concerned, an identical process can be observed, although with taking local variations into consideration. In the Carpathian basin groups of 5 to 20 deceased buried in the settlement's area, part of those often being children, were noted[2]. When focusing on Danubian societies, most researchers agree that the first cemeteries appeared in the late stages of this culture, although some of them oppose this view. One of them is the cemetery in Vedrovice which is dated at a very early period (5,500–5,300 BC)[3]. 12 graves located in the settlement's area were found there, out of which 9 were child burials and three were adult ones.

---

[1] Chapman 2000, 168.
[2] Lichter 2001, 263.
[3] Podborsky et al. 2002, 317.

Regular burials in settlements remained a characteristic feature of Danubian societies throughout their whole development. According to Midgley, burying the dead in settlements during the Neolithic period in Central Europe refers to the traditions of Mesolithic hunters groups, according to which celebrations, feasts and other unknown ritualistic activities were performed, often with the use of fire over the graves[4]. As Midgley suggests, it would be justified to see a transformation period throughout the whole Neolithic period in Central and Northern Europe, when the Mesolithic and Neolithic traditions mixed with each other, which is confirmed by additional facts: Danubian style stone tools found in the hunter–gatherer areas and the opposite–Mesolithic amber and bone jewellery found in Neolithic settlements. Some researchers see a source of another stage of farming societies in this early period of mutual relations, which was characterized by inter alia the monumentality of tomb architecture, of which I wrote about in the previous chapter[5]. The synthesis of both traditions may be also sought in the hunter/warrior tombs described by Gronenborn[6], and even–on the processual level–in the origins of the traditional depositing of axes in chosen locations by Mesolithic hunter–gatherers[7].

The process of separating the dead from the living proceeded gradually, which shows just how strong the tradition of keeping close relations with the deceased was. Abandoning this custom signalled a stronger symbolic separation of these two spheres. In fact this separation also meant the denial of monumentality as, according to Chapman[8], it was connected with tells or long houses. However, what is interesting is that with liberating the dead from the dominating order of the settlement or house, orienting them according to cardinal directions became very important, which at the same time replaced the architectural model applied to the burial of their ancestors. Such a situation has been confirmed at many Neolithic cemeteries throughout Europe, both in its South–East part (Durankulak, Poljanica, and Varna) as well as in Central–European cemeteries. Depending on the specific culture and time, the different directions of placing bodies of the deceased were applied, although analysts suggest that a lot of effort was put into performing it as accurately as possible. In the south of Europe the North South line direction was most often used, with small deviations. However, further in the North, in the

---

[4] Midgley 2005, 65.
[5] Zvelebil 2005, 87–101.
[6] Gronenborn 2003, 35–48.
[7] Koch 1998; Migdley 2005, 45.
[8] Chapman 2000, 156.

Carpathian basin, the East–West direction became most widespread and was used throughout the Eneolithic[9].

The problem of orienting the deceased had been given more attention at Vedrovice, in Moravia. The graves discovered there were placed in two basic configurations. Most of the deceased were laid with their heads towards the South–West. A small number of graves were oriented in a different direction–heads towards the South. When conducting subsequent analyses, the study's authors concluded that most of the graves were probably connected with observing the Sun, and more specifically its rise during midwinter. This is quite surprising as it seems that grave orientation was not connected with the farming cycle, when its presence in farming societies seemed obvious. Another group of graves would be connected with the summer lunar movement, which seems even odder when taking into consideration how difficult it is to correlate the complex lunar movements. However, a part of the graves did not point towards any astronomic allocation[10].

The analyses of gender regarding astronomy brought up some interesting results. It appeared that men and women were oriented completely different than children. Dead children were oriented towards the Sun's rises. A theory has been formed according to which child burials were probably connected with the belief in the so called "new sun"–an idea of the birth of a new life with the start of a new solar cycle[11]. When summarizing the studies it has been concluded that the cemetery's placement had been chosen to correspond with cult–astronomical requirements. It was supposed to be a type of a cult calendar which obeyed the rhythm of the Sun's rises and sets as well as the observations of the Moon's full periods. Numerous astronomic correlations have also been observed in Bavarian cemeteries. Most of the deceased (above 80%) were oriented in a NO–SW and SO–NW line[12].

It seems that during this process of separating cemeteries from settlements a discourse connected with the different treatment of gender in graves had occurred. The diversification of burial rituals in terms of gender appeared first in graves of the early Copper Age in the Balkans[13]. Placing men in graves in a straight position on their back and women in a curled position on their right side was characteristic at that time. Graves were oriented in a North–South line, sometimes with some deviations

---

[9] Lichter 2001.
[10] Rajchl 2002.
[11] Ibidem, 275–291.
[12] Nieszery 1995, 69.
[13] Lichter 2001, 72–74.

from this direction. Although male graves were usually more richly supplied than female ones (these were stone axes, later on copper axes, spondylus shells, marble vessels), typical male–female accessories, like in the case of late Eneolithic, cannot be precisely separated. It must be mentioned here that the supine position is seen by many researchers as the remnants of Mesolithic hunter traditions[14].

The differences in gender became all the more clear in a later period and another place which was connected with the Copper Age of the Carpatian Basin. In the Tiszapolgár societies men were placed in graves in a curled position on their right side and women on their left side. The items placed in the grave also differed. Male graves usually contained e.g. stone, copper and bone axes, boar tusks, flint cores and blades as well as pig jawbones, whereas what archaeologists discovered here were decorations made of stone or copper beads. In the oldest stages men were also placed in a slightly curled position[15]. Finally, in Europe there appeared certain societies in which the male–female dichotomy and its geographic–astronomic direction started to be one of the main distinguishing features of the culture discourse. We shall speak of this later on.

There is no doubt that the above mentioned socio–cultural transformations created an essential level of forming a new ideology, in which framework objects acquire an ultimately different value from the previous one. As I showed in chapter 1, the problem of human–object relations is the key for this argument. The early Neolithic/Eneolithic period was characterized by emphasizing the narrative value of objects connected with symbolic and ritual value. The most characteristic of these were objects made from the spondylus shell, a shell which may be found in the Mediterranean Sea. Objects which had been spread throughout the whole of Europe occupied by early farming societies were made from them. The attachment to spondylus shells was present both in Neolithic and Eneolithic societies, which shows a strong and long–lasting tradition. It was undoubtedly a very valuable material as it is discovered in richly supplied graves. Statistically speaking only 10–15% of graves were supplied with objects made from these shells. This is not a great amount, which presents the shells as being and accessible only to the most respected individuals[16].

The spondylus shell's circulation system is often compared to the kula exchange described by Bronisław Malinowski. His research showed that

---

[14] Häusler 1981, 101–149; 2007, 55–77.
[15] Ibidem.
[16] Müller et al. 1996, 81–96.

the exchange between individual partners was performed in short distances as the shell could not be stored for long. The narrative traditions, which are the myths and stories of a given society in which the spondylus objects were present, was probably focused around a strict, ceremonial exchange. Objects in such a system did not belong to anyone in particular, they circulated in a seemingly eternal exchange process, during which they gained a specific symbolic value of a narrative character[17]. Moreover, the formal richness of the shell objects is very great which means they were not only exchanged but also altered, constantly changing their form, even the small beads could only be used to create necklaces. Hence, in the wider perspective the spread of shells had to cover a large area, as it is actually observed in the archaeological material. The first archaeologist who considered such an interpretation was Colin Renfrew[18]. He stressed that one of the probable aims of acquiring the spondylus was to emphasize the relation with areas of an older Neolithic tradition from the south of Europe.

We cannot say much of the myths, which accompanied the ceremonial exchange of the spondylus. However, we may imagine them through ethnologic analogies. More light on this matter is shed by such singular discoveries as the one described by Seferiades. An originally ornamented pendant made from a spondylus shell was found in one of the Neolithic graves in Voivodina on which, according to the researchers, fish, stars in the sky and a house placed on posts were engraved. According to Seferiades this representation can be seen as a mythogram, although we cannot say more about the deeper message it bears[19].

In the context of spondylus ornaments' production and exchange we approach a fundamental phenomenon of the European Neolithic/Eneolithic period which is intentional fragmentation. These objects were strictly linked with a wide range of symbolic activities–the fragmentation and deposit treatments, which were confirmed by Chapman, Gaydarska and Slavchev when studying such sites/cemeteries as Varna and Durankulak in Bulgaria. Many shells were deposited in graves as whole objects, some graves have a distinctively large amount of these, but there were also a lot of shell fragments deposited both in as well as outside graves. This image is completed with the custom of replenishing the shells fragmented earlier with new fragments by installing a sort of filling[20].

---

[17] Dzbyński 2008, 89–106.
[18] Renfrew 1976.
[19] Seferiades 2000, 423–437.
[20] Chapman et al. 2008, 139–162.

Researchers conclude that the enchainment of human relations by the use of fragmentation was a basic social practice aimed at symbolically connecting the domains of the living/life with the domain of death. Personal and group relations towards the deceased were cultivated by storing the fragments of shell ornaments in the form of rings, where the remaining parts were placed inside graves in Varna and Durankulak. Moreover, the spread of the spondylus objects' fragments on greater areas should be connected with the lack of these in numerous graves. According to researchers, relations with the landscape were also made in this fashion. Some of the ornaments were scattered in places significant for the deceased, while the chain's end symbolized the other fragment deposited in a specific grave. The majority of the found spondylus objects are in fact more or less fragmented.

Spondylus ornaments undoubtedly carried a prestigious value and were connected with "wealth" which is confirmed by such items as inter alia a shell bracelet with golden additions, which was found in one of the graves at the cemetery in Varna. This example shows a material connection of old narrative images with metallurgy technology, which began to have a great influence on people.

The enigma of the first axes

Studies focused on large stone tools are equally interesting. Stone axes became very important tools in the social life of the Neolithic and especially in the Eneolithic. They were used intensively, also as weapons; however the evidence of their symbolic significance is their constant presence in graves as well as being deposited in large numbers. Moreover, the axes discovered in some graves were made from an unusable rock material, which was too fragile[21]. Among the first farming societies in central Europe (Linear Pottery Culture) axes were put into the graves of 30–40% of the society. Most often these were adult (30–50 years old) or older men, almost never women[22]. Bakels thinks that perhaps the axe was a taboo for women. At the cemetery in Nitra, Slovakia, it has been estimated that older men, more than 50 years old, had an 80% probability of being buried with a stone axe. Among other objects which often accompanied axes were flint bolts, used as tips for arrows, vessels and spondylus ornaments[23].

---

[21] Bakels 1987, 156–157.
[22] Ibidem.
[23] Pavuk 1972, 5–106.

Some graves contained more than one axe. Most often of two types: one large and one smaller light weight one. It seems that the purpose of this was not only to symbolize different activities connected with axes as such but also to emphasize the status of the buried individual. Did different types of axes also represent different social values?

Hans–Dietrich Kahlke recently presented an interesting theory concerning the development of these early polished stone tools[24]. He wondered about the significant morphologic diversity of these tools in "linear" cultures. There are different known types of axes and stone objects used for similar functions to those of an axe, which could have served more specialized tasks. Archaeologists give these various names: lasts, hoes, stone cleavers, but the marks found on them undoubtedly point to the fact that they were used to work on wood, which means they were basically different types of axes and chisels. Basing on this fact, the thesis according to which some of them were used for earth work in the field as a type of hoe has been dismissed. Simple wooden tools were probably used for this purpose instead of stone ones[25].

While studying objects from different cemeteries, Kahlke noticed that such a diversity increased in time: the forms of axes, chisels and lasts became more and more distinct. When studying objects from the early stages of this culture it was often impossible to clearly guess what type of tools we are faced with. Hence, Kahlke brought up a thesis according to which the earliest type of stone tools in the first central European farmers' culture were tools that were cross–section shaped somewhat like axes, which nonetheless started to evolve quite rapidly into two basic directions. Some became flatter with a broad blade, others became longer or, to be more precise, thinner. Included in this group are both very long specimens, which researchers definitely interpret as non–functional[26], and miniature chisels. In the earlier stages of these societies we may distinguish three basic tools, out of which only one kept the form of a true axe.

According to Kahlke, there had initially been only one type of tool which was used as an axe and a symbol of the male social role. This interesting hypothesis lets us see that the stone axes were socially important tools. The developing formal diversification into tools of a probably more specialized function lets us attach to it the growing social diversity of the first farmers, the increasing urge to emphasize the emerging individuality. This had been done by manipulating the form of this very important stone tool.

---

[24] Kahlke 2004, 90–94.
[25] Kruk & Milisauskas 1999, 42.
[26] Weiner 2003, 423–440.

An appropriate image of the social reflection of the axes' morphologic differentiation mentioned above is in the cemetery in Niedermerz, North Germany[27]. Here too did some of the men have a stone axe put into their grave. It is also the place where two basic forms of axes were present: a large, long one and a slightly thinner and flatter one. The distribution of these two types of axes in the graves in Niedermerz shows that they were subject to different symbolizations and valuations as in each of the two groups of graves, which were located slightly further from each other; they were placed only in those on one side of the cemetery. The graves with different types of axes were not in close proximity to one another. Not only would this confirm their different tool character but also a different social valorization. Moreover, in each of these groups there was one burial in which both types of axes were put into the grave, but only in two centrally situated graves.

The description presented so far allows us to come to a conclusion concerning the stress put on the form of signaling social differences among the first farming societies. It seems to me that both the spondylus objects, which carried various narrative content, as well as the prestigious stone objects could not yet have been perceived in the metrologic–numeric categories, hence such distinct stress was put on their formal differentiation. In the case of spondylus objects–whole shells were used and not just fragments. Hence, it is probable that a strong trend was present in the society which led to symbolization in the form of concrete shapes and images, similarly as it was in the case among hunters and gatherers[28]. Numbers could not have stood behind symbols but only narrative values, which were hard to quantify. As we shall see below, it was only with the development of Eneolithic flint tools that a distinct change occurred. Their shape remained the same, more or less, but the stress had slowly been put on an abstracted tool value in the form of their length (Fig. 4.1). In my opinion it is evidence of forming a new way of representing the measuring stick metaphor on which I shall elaborate below.

---

[27] Dohrn–Ihmig 1983, 47–190.
[28] Wierciński 1994, 53–69.

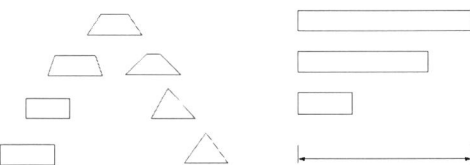

Fig. 4.1 In the Linear Pottery Culture axes were valorized by putting emphasis on their morphology (on the left), while among the later Eneolithic cultures the emphasis had been moved to the tools' length (on the right).

Axes in the Eneolithic period

A uniformization of the axe occurred in the Eneolithic period. It gained a clear quadrilateral cross–section, a trapezoid shape, a straight blade and a clearly distinguished head, although there are some exceptions to this rule, especially when considering axes made from rare rocks. Moreover, there had been a focus on producing flint axes, which is sometimes defined, along with the fashion for long flint blades which spread in time, as the macrolithization of flint tools[29]. Let us first look at axes.

The production and distribution of flint objects, including a large amount of axes, was based on the existence of a network of settlements. Groups of miners and stonemasons went to a quarry where raw material had been mined and initial processing occurred. Most often the initially processed material, in the form of the so called rough–outs, were brought to the settlement and made into axes or other tools in the workshops located in the part of the settlement prepared for such tasks. Finished tools were afterwards exported to settlements further from the flint outcrop. The ranges of distribution differed and depended on the discussed period and the material it concerned[30].

Flint axe production was mixed into a vast network of associations. It seems that the development of this manufacturing branch is also connected with a more rational use of flint material by specialized flint tool makers. In regions containing rich flint deposits, worn out cores used for producing tools were often treated by earlier Neolithic societies as scraps, rarely

---

[29] Dzieduszycka–Machnikowa 1961, 29–31; Lech 1991, 557–569.
[30] Petrequin et al. 1998, 277–312.

using them in households as pestles or mallets. However, in later years such cores became a valuable source of material for making stone axes[31].

Bogdan Balcer made an interesting observation. After overlaying outlines of several dozens of such axes he perceived something close to a metrological axe standard (Fig. 4.2). Prehistorical craftsmen aimed at making "new" flint axes more or less fit some set sizes. This observation was left unnoticed by archaeologists because of the lack of a proper theoretical level. However, it clearly points to the fact that at that time a measure standard could have existed and used during axe manufacturing. Although this observation alone does not imply that these objects were perceived in a metrological/numerical fashion, the examples presented later on shall allow us to support this point of view[32].

In time, the manufacture of flint axes deviated more and more from a utilitarian character. Environmental changes, the gradual transformation of some areas of central Europe into steppes as well as the processes of the decreasing fertility of soils extensively cultivated (through burning large forest areas) by Eneolithic societies opened up new possibilities for the economy and lifestyle of the people. Some researchers think that the main reason for that period's changes, connected with the crisis of the former way of life, are in fact the anthropological changes of the natural environment, which limited making further use of slash and burn agriculture. The environmental changes imposed some changes in culture. An increase in the breeding of animals for farming occurred, which led to corresponding social transformations such as the gradual independence of animal breeders, who introduced new social rules and norms, new symbols etc.[33].

We can guess that, in the above mentioned context of socio–economical changes, the flint axe had to function in a slightly different symbolic dimension. In theory, societies which put a greater focus on breeding animals than the cultivation of crops did not show any greater interest in this tool, as it is not required for so many tasks as in the case of activities connected strictly with farming. However, the meaning of the flint axe in some European regions increased, which is clearly seen in the context of the proper mining methods of extracting flint, the great number of axes, and the symbolization and numerous mysterious manipulations

---

[31] Balcer 1983.
[32] Some archaeologists counter his argument stating that small axes had also been made, which do not fit the metrological standard mentioned here. However, this is not a valid argument. Small axes could still have been manufactured with relations to the standard (compare Kripke's arguments from chapter 1).
[33] Kruk & Milisauskas 1989, 77–96.

connected with it. In that period the extraction of flint developed in a way never seen before.

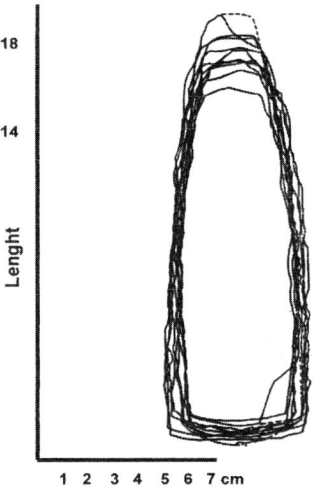

Fig. 4.2 The sizes of Eneolithic axes according to Balcer 1983.

Scandinavian experiences

The experiments conducted by Dutch researchers show that flint axes could have been effective tools, which probably also constituted to their value [34]. However, axes were not used for their seemingly obvious function, which was for chopping down trees and clearing forests. In other words, they were efficient tools but apparently their functions were limited. During experiments flint axes broke very quickly when used to chop down trees. They most often broke as an effect of the high tension caused when the angle at which the axe hit the tree was too great. However, such a situation could never be foreseen or prevented; hence such an axe could be broken at any time. Another mystery was the flaws of the material from which the axe was made. Limestone inclusions often do appear in flint materials, which make it less durable. While on the outside everything seems fine, the flaw is present inside the material. These factors caused a lot of uncertainty when working with flint axes. One could never be sure if the axe would crack or break in half making further work impossible. It is quite certain that these factors were perfectly known

---

[34] Olausson 1983, 1–70; 1997, 269–278.

to the axe users of the time. It is hard to imagine a man chopping down trees and working effectively using such tools, even if he were equipped with a couple of them (quite impractical) after venturing into a forest. It would still be insufficient. Environmental studies seem to confirm that the most common method used for clearing fields and pastures for cultivation in the Eneolithic period was by fire. In addition, the ash created as a result was an excellent fertilizer in times when animal manure was not used[35].

Hence, it is believed that axes were better used for working near houses, for example for carpentry or simply to chop firewood. Hitting wood perpendicularly with an axe, which it was perfectly suited for, was much safer than hitting it at an angle. Because such actions were everyday chores for Eneolithic people, axes were often used on farms. However, if we assume that they were also a very valuable tool made by specialists, often brought to a village from far away, and then using them in a homestead was also a way to present oneself. An Eneolithic farmer who chopped firewood on his own farm could proudly display the valuable object to his neighbours.

As flint axes found in the Danish Islands were characteristic for their significant length, there have been suggestions that their metrical difference could have had some influence on the value or function of axes. What stimulates us to perform such an analysis are ethnological observations. Various relationships have been noticed within this framework. Again the point of reference has been the work of Bronisław Malinowski. He observed that basalt axes had been used on the Triobrand Islands, whose deposits were situated about 100 km from the settlement. This material carried aesthetic attributes for the island's natives. Some axes even had a special trimming, which raised their value. A man who possessed a particularly large axe or several small ones was granted social prestige. He could exchange them for other goods or services, enter marriage, etc. The chief was always responsible for bringing axes to the village and for their distribution, which is why he regularly sent gifts to specialists in processing basalt into axes. Nevertheless, in time the most impressive pieces became a sort of heritage passed down among the clan that the owner was a part of and, as a common good, presented during different rituals, and sometimes even hidden in a forest for fear of it being stolen[36].

In another case, observed in New Guinea, stone axes were used on different occasions. There was no advanced specialization and basically

---

[35] Kruk & Milisauskas 1999, 403–446.
[36] Olausson 1983, 12.

each man, after receiving proper training from elders, could make his own axe from one of the two available stone outcrops. Moreover, these axes could have been subject to valuation and classification; their three basic functions had been distinguished as axes for working, axes for buying a wife and ceremonial axes. The natives themselves could not put an axe into any one category on the basis of its shape and workmanship. It seems, however, that their functionality had some influence as sometimes axes classified as being ceremonial had been used for menial house tasks. On the other hand, everyday–use axes were used for completing any other tasks concerning the house or the farm and every man performed about half of his everyday tasks with an axe[37].

However, ceremonial axes could not have been used for hard tasks because of their quality. A proper ceremonial axe was made with great care, polished, often made from a colourful, eye–pleasing stone, 27–30 cm long and only 0.8–1.5 cm thick. Ceremonial axes were also used as weapons and carried a prestigious value. It sometimes took a whole week to make such an axe[38]. The largest axes reached 45cm in length and were used as a marital gift. Their size excluded any practical use. However, it was not enough to simply own a single huge axe in order to get married. One had to acquire 20 or 30 of them! Hence, it is no surprise that the production of such axes, the process of their collection, sometimes by buying them from other members of the society, was a very long and complex process which influenced the shaping of social relations[39].

At first glance, the ethnological examples seem to ease the process of differentiating ordinary and functional from ceremonial and prestigious axes present in archaeological material, although Olausson showed, via an experiment, that the length of an axe does not disqualify it as a practical tool nor did it change significantly along with the distance from the sources of the material[40]. However, axes were connected with ritual depositing on many levels.

First of all, what stands out is the great diversity of the sizes of Danish axes. The size range of these objects ranges from 13cm to over 40cm and the first comparisons did not shed much light on the issue. It appeared that axes of different sizes had been put with other artefacts. Despite expecting that the depositing of axes in the ground was aimed at burying the most valuable (whether according to economic, religious or social reasons), there are also medium and even small ones among these artefacts.

---

[37] Vial 1940, 158–163.
[38] Olausson 1983, 12; compare Stout 2002, 693–722.
[39] Petrequin et al. 1998, 295.
[40] Olausson 1983, 9 and n.

Therefore, depositing axes could not have been perceived only as a wish to hide or ritually bury the best axes. However, it could have been a practical procedure of storing a supply of tools for some other time[41].

Two to three axes have been found in most of the deposits, while finding more than 8 in one place is extremely rare[42]. However, all types of artefacts, both humble and the very rich are in general spread evenly across the whole area of the Danish Islands that were populated in the Eneolithic period. Their even spread, whose points can be connected by lines, suggests that these deposits were often situated along communication routes. Most of these deposits are also found in close proximity to megalithic tombs, although axes were rarely placed inside them (only a couple of short ones were found). Megalithic tombs, as we remember, were built in exposed places on purpose as they symbolized the claim of the local society over a particular area. Ancestors were buried inside them, who confirmed the claims to the area and were visible to outsiders. Hence the circulation of flint axes, travel for exchange purposes as well as the mythological–ritual sphere connected with constructing stone tombs were all mixed into one complex network of relationships. The fact that axes became objects filled with deep symbolism is visible in the numerous charms of that period (Fig. 4.3).

Kristiansen thinks that a very strong connection between the axe and all forms of social activity had to exist among megalithic societies. It became a basic reference plane in tool manufacture, deposit ritual, exchange and the tomb context. There probably existed a closed cycle of social activity in which the axe played a central role. Finished axes were used as a means of exchange, ritual consumption and they were necessary for upholding the progress of farming production (as a tool required for deforestation, even if only symbolic)[43].

Nielsen's analyses are the representation of the close relationship between flint axes with ancestors and megalithic tombs. The mass depositing of axes in Denmark was initiated with the appearance of the first ground graves. The intensity of depositing increased further in the next stage when stone megalithic tombs started to be constructed. It ceased along with the introduction of passage graves, which could contain many more burials and at the same time emphasized community rule[44]. On the other hand, the early dolmens present on the isles, which were characteristic for their closed tomb chambers, are again connected with the

---

[41] Ibidem.
[42] Rech 1979, 19.
[43] Kristiansen 1984, 79.
[44] Nielsen 1977, 61–138.

intensive distribution and deposit of long flint axes. Hence, the exchange and deposit of axes was strictly connected with the concepts concerning megalithic tombs, they functioned as an element of a common discourse and a network of metaphors, although at present it is impossible to interpret this phenomenon without some doubts.

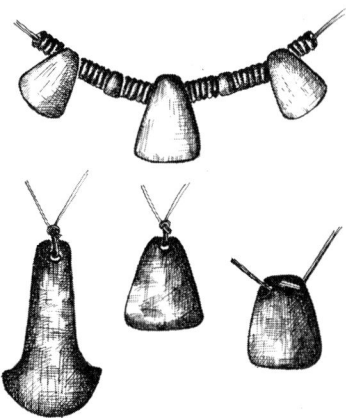

Fig. 4.3 Eneolithic axe–shaped pendants and charms
(according to different authors)

Based on the facts mentioned above, can we expect a credible transfer of observations of the archaeological evidence on the sphere of vocabulary and metaphors in the Eneolithic period? When studying this matter, Whittle mentioned the differentiation between the metonymy (part as a whole) and a metaphor (arbitrary association)[45]. By referring to Leach[46] in this way, he imagines that the axes could have been used metonymically as a symbol of a much greater sphere of managing: occupation cycles, cooperation between widespread farming societies etc. In this context he calls upon the example of menhirs, which were sometimes formed in the shape of gigantic axes. Whittle suggests that their metonymically monumental size was a metaphor of ruling over nature.

Davies shares this opinion. According to him the menhir shape itself could raise similar emotions as those connected with axes[47]. If axes could be used for presenting social or gender status then why should menhirs not

---

[45] Whittle 1995, 247–259.
[46] Leach 1976.
[47] Davies 2012, 139.

play the same role? Tilley adds that the relationship between menhirs and axes could also refer to the role played by axes in taking control of the forest[48], although we do remember that chopping down trees by using axes was quite a troublesome process. Moreover, Tacon supplements this by stating that ethnographic notes show that axes often symbolize male strength, penetration and power[49]. As I shall try to show below, this does not cover all the metaphors connected with the axe. This is best proven by the example of the Eneolithic Globular Amphora Culture.

Striped flint

What is not yet fully apparent in the case of Danish axes becomes clear in the production of striped flint axes in the area of modern Poland. Let us begin from the source of this phenomena, which are the mines in Krzemionki Opatowskie, one of the largest flint deposits in Europe. Today we can visit the prehistoric underground corridors and witness at firsthand which mining methods were used more than 5,000 years ago.

Flint deposits were mined in Krzemionki Opatowskie from all types of rocks: from loose ones, which were nearest to the surface, to monolithic rocks, which were up to 8 meters below ground. In the eastern part of the mining field, where flint deposits are situated 2 meters below ground, straight circular shafts 3–5 meters in diameter were dug. However, the most characteristic feature of Krzemionki was the practice of digging shafts and underground headings 8 meters deep into limestone. These were enormous exploitation units which provided large amounts of flint. The rooms were connected to each other creating an impression of a system of walkways. Pillars were set up in order to prevent the ceilings from collapsing. Very narrow corridors were used to move through the footpath laid out with limestone slabs. Researchers call this way of mining as room–and–pillar mining. The height of the underground headings ranged between 55cm and 110cm. Because of this, prehistoric miners had to move and work on their knees or in half–lying positions[50].

Miners were strictly connected with the life of nearby settlements, from which they received food, while half–products made in the mines were transported to the settlements. The estimates made by researchers studying Krzemionki show that the amount of axe production was significant and could range from dozens to hundreds of axes per year[51].

---

[48] Tilley 1996, 113.
[49] Tacon 1991, 204.
[50] Borkowski 1995.
[51] Balcer 2002, 152.

The labour required to make an axe was divided into stages: mining materials and initial processing were both made in workshops near the mines, while the most labour-intensive stages: precise processing, sanding and polishing axes were performed in the manufacturers' settlement, who not only made new axes but probably also repaired broken tools.

Finished striped flint axes spread to settlements far and wide, even up to 600 km from the mines. The carefully sanded ones became standard parts of burials performed by Eneolithic Globular Amphora societies in the area of today's Poland and Ukraine. However, the largest and best–kept regional concentration of those is found in the Kuyavia area, where they were deposited very often in megalithic tombs, opposite to what was happening in the Danish Islands. Kuyavia may be treated as a region in which, at the time of the Globular Amphora societies' settling, a circulation of information common for that region was present, which had an influence on its separateness[52]. The analyses of the settlements allow us to assume that the activities of individual local groups took place in a limited area and had a connection to locating megaliths. Societies of the Globular Amphora Culture farmed in close proximity to the tombs connected with them, at the same time pointing to their attachment to the territory, which was both their area of exchange activities as well as of symbolic value[53].

Researchers were right to underline the small functionality of striped flint axes. Similar to Danish axes, they were vulnerable to breaking or cracking during use, and so it is possible that they were also used for less strenuous activities. Moreover, the careful sanding of the whole axe had no influence on its usefulness; it only increased the amount of labour required as well as aesthetic values. Earlier flint axes, connected with the Funnel Beaker societies, were sanded only by the blade, which may be understood as an attempt to improve its usefulness. However, axes were made from different materials than striped flint in the beaker culture period and their existence covered a much smaller area[54].

After performing more systematic analyses it appeared that even the precisely made axes showed signs of modifications and repair. Using the statistic morphological description method (all features of the shape) on flint axes made from striped flint, archaeologists from the National Museum of Archaeology in Warsaw discovered that they show numerous

---

[52] Szmyt 1996.
[53] Wiślański 1969.
[54] Balcer 2002; Borkowski & Migal, 1996, 141–165.

signs of work and modifications. Approximately 65% of these objects bore clear signs of being repaired[55].

After taking into consideration the above mentioned information concerning these axes I decided to focus on the matter of the decreasing size of axes. For example, it appeared that the size of flint axes, which were put into graves with the deceased in Kuyavia, was not constant and depended on the localization of the grave. Large axes are most often discovered at richly equipped cemeteries which were used for a long time and around which other graves are located. On the other hand, small axes dominate in peripheral areas in smaller cemeteries where there were few graves (Fig. 4.4). Therefore it seems that in this case we may perceive the connection between the size of the axe and the symbolic dimension, as the size of the axe was most probably significant for the valorization connected with the localization of the megalithic tomb, which allows us to assume that axes as such were subject to some kind of measurement and classification in Globular Amphora societies in Kuyavia. This assumption has been confirmed by statistical analyses which took into account the localization of the graves and the size of the axes found there[56].

The fact that the size of the axes were subject to such a valorization also finds its confirmation in the analyses of the size of axes from graves in which the gender and general age of the deceased has been identified. For it appeared that in mass burials, during which women and children were buried, axes are significantly smaller than in male graves. Although this may be explained by the fact that adult men were entitled to have larger axes, this does not in any way diminish the hypothesis, according to which axes were valorized because of their size, more precisely their length[57].

In the light of the above mentioned analyses we may assume that new striped flint axes left the workshop area of the mines in Krzemionki in the form of metrically standardized objects. The earlier analyses preformed by Balcer concerning the Globular Amphora Culture already showed that the "new" flint axes made at the source of the material had roughly similar sizes (Fig. 4.2). We may set this size at being between 15cm and 18cm, although a few examples of a slightly greater size have been noted. During its use, an axe was subject to wearing from the tension present during work or as a result of other effects or events. As experimental research shows, the damage which occurred most often was the cracking of the

---

[55] Borkowski & Migal 1996.
[56] Dzbyński 2008, 76–80.
[57] Ibidem, 81.

blade across the axe[58]. The axe head could also break off as a result of the loosening of the hilt in the handle. As each new axe could have been repaired only a specific number of times (no more than three to four), their length was physical evidence of the number of such repairs. As the length was a constant component of ritual practices we may assume that the whole process of modification was an important point of social focus. Mounting the axe onto a hilt did not necessarily limit the possibilities of assessing the axe's length[59]. It may be said that each of the flint axes had a sort of biography which, apart from its narrative value, was also capable of protonumerical description, which I would like to consequently define as another manifestation of the measuring stick metaphor. Analyses show that it was also the basis for placing different size categories into different social contexts connected either with the localization of megalithic tombs or with the gender, age or social status of the deceased.

Another interesting fact is that in the case of the Kuyavia Globular Amphora societies a long lasting lack of copper objects, which were an increasingly common material of use among Eneolithic societies in Central Europe has been noted. For example, in Central Germany traces of intensive metal processing, which accompanied the development of sepulchral styles and architecture of megalithic origin, has been noted which is also explained as a reaction of metal expressed by means of a ritual[60]. From this perspective, the lack of metal and a reasonably small architectural modification of tombs in the Globular Amphora societies in the area of modern Poland in fact point to some sort of conservatism. Here the exchange of striped flint axes was probably one of the mediums of expressing individualized social relations based on the simple conceptualization of number/measure. However, flint technology became a dead end as it blocked the development of more abstract mathematical concepts.

---

[58] Olausson 1983, 51.
[59] It must be mentioned that small axe heads, those less than 8cm, could simply not be fit for mounting.
[60] Klassen 2001, 300.

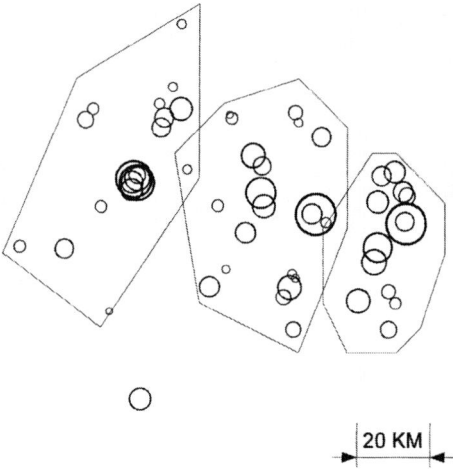

Fig. 4.4 Analysis of the dispersion of axe sizes in megalithic graves from Kuyavia (after Dzbyński 2008).

Not only flint

We shall end the study concerning axes with the research work results of Czech researchers. In the Eneolithic period in the area of the Czech Republic the material used for axe making was not flint but a characteristic green–coloured stone–volcanic tuff. And so by starting from the question concerning the axe's usefulness we can reach the conclusion that it was not at all useful, as tuff is a rock which is too soft to be used as a practical tool. On the other hand it was eye–pleasing and was easy to process. It seems that this is how we can explain the local Eneolithic societies' interest in this material. It was only symbolic, aesthetic and non–functional.

In his study concerning the distribution and use of stone materials in the Eneolithic period, Jan Turek discusses the collection of sanded stone axes made from green volcanic tuff at the Řivnac sites, which existed there in 3,300–2,900 BC. According to him it was the characteristic colour of the tuff which caught the attention of the Eneolithic people. Most of these axes were found inside enclosures, of which I wrote about in chapter 3. At some sites they constitute 38–96% of the whole material discovered there, which is a very high percentage. However, this material has different estimates at sites of the neighbouring Cham Culture, where tuff axes constitute up to 15% of the whole collection. The situation is similar as far as the distribution of other categories of findings is concerned, which are connected to these two culture groups. While at the Řivnac sites traces of all production stages of these tools can be found, only the finished objects are noted in the Cham sites[61].

The statistical analysis of the small axes performed by Turek distinguished two basic sizes, which in fact have an equal share in both of these cultural groups. According to Turek, mostly small axes were adapted to the Cham societies' area in the form of finished objects. Meanwhile, at the Řivnac societies' sites a greater variety of imported objects were noticed. Hence, he concludes that the production of the green crystalline tuff was most probably controlled by the Řivnac societies and the exchange of axes between both regions was a form of ritual communication.

Macrolithic blades

If despite the arguments presented above the implementation of mathematical thinking by the use of production, exchange and depositing of axes in prehistoric societies still seems elusive to some readers, the case of macrolithic blades should discard this impression.

The first macrolithic blades were produced in the area of today's Greece around 7,000 BC. According to researchers it was quite an odd innovation which remained a local phenomenon[62]. These are so rarely found among Eneolithic South European cultures (the blade's length reaching up to 15cm)[63] that any analyses concerning their social meaning are difficult to perform. Perhaps this is a result of an intensive interest in the obsidian rock of those times, which was a material not suitable for long blade manufacture[64]. The most intriguing macrolithic blades appeared again in the areas of the Chalkolithic societies that populated South–east Europe, where we are first confronted with the phenomenon of accumulating large amounts of metal. In the richest graves of those times several kilos of copper and gold have been found in the form of such tools as axes, chisels, cleavers as well as ornaments such as gold plates of various shapes, pauldrons, necklaces, and gold buttons etc. In the context of this richness, which appears suddenly in European prehistory, long,

---

[61] Turek 2001, 53–62.
[62] Pellegrin 2006, 37–68.
[63] Kozłowski & Kaczanowska 1990.
[64] Kaczanowska & Kozłowski 2008, 9–37; Bácskay & Siman 1987.

regular flint blades, undoubtedly products of advanced craftsmanship always had their place among other artefacts[65].

According to the analyses performed so far by researchers studying the Copper Age in Bulgaria, the dimension of a blade was definitely important. Laurence Manolakakis' research illustrates this in an exceptionally clear way: the richer the burial, the greater the length of the blade in the grave. The evaluation of richness in the Varna graves is all the more easy as in the richest of them several kilograms of copper and gold in the form of different objects had been placed. On the other hand, the fragments of identical blades were found in less equipped, or simply poorer graves. When taking into consideration this fact we may with all certainty assume that they were a very important status symbol (similarly to copper), symbolizing wealth and prestige[66].

Slightly smaller blades, reaching up to 25cm in length, were produced in Central Europe. They are not as fine and sleek as the Balkan blades, but more importantly, as we can undoubtedly deduce, macrolithic blades in central areas of Europe had their practical use as many carry signs of wearing in the form of harvest wear–out. The best known cemeteries, and at the same time largest, of that period are Tiszapolgár–Basatanya in Hungary and Tibava in East Slovakia. The burial ritual at these cemeteries was clearly determined by the dichotomy of the deceased's gender. Men were placed in a slightly curled position on their right side while women on the left, each gender was faced in opposite directions. Apart from such obvious differences other features of burial ceremonies have also been noted. In male graves one can find characteristic objects such as copper and stone cleavers as well as stone axes. However, these are not common objects: copper cleavers are found in only 16 male graves while axes are found more often[67]. Copper and gold ornaments appear in both male and female graves.

Despite the clear analogies between both areas the (a dynamic development of metallurgy, copper axes and cleavers present in graves), macrolithic blades present in the graves of Central European societies were subject to completely different rules of leaving objects in graves than those discovered in Bulgarian graves. It appears that in the context of the best supplied graves blades are rather small and fragmented to a significant degree. Only about a half of them are long blades, those above 10cm in length, although it has to be admitted that this size is not significant when compared to the Balkan blades, which reach up to 40cm.

---

[65] Manolakakis 2005; Tsonev 2004, 17–23.
[66] Manolakakis 2005, 230.
[67] Lichter 2001, 282.

For example, in one of the graves in Tibava, Slovakia, 25 fragments of blades of different length were found. It is also one of the richest graves in the whole of the described society, as it also contained objects made from gold, a copper axe, copper bracelets and other objects[68]. Hence, it is a completely different situation than that of the Balkan cemeteries.

As to the Tiszapolgár cemeteries in Slovakia, one more interesting element should be noted. Among the 41 graves in Tibava there were 7 in which large unprocessed flint nodules had been placed. On the basis of the graves contents, these graves should definitely be counted as rich or very rich even. All of them were supplied with a large number of flint vessels and blades of different lengths, copper axes or copper wire cords and even golden pendants. An identical situation was observed at the cemetery in Vel'ké Raškovce where 6 out of the 44 graves explored there contained large flint nodules, as well as very rich items[69]. The flint nodules came from Volhynia and presented an undoubtedly great value as it was a type of flint very sought after at that time in the south–in the Carpathian Basin which was the copper manufacturing centre of that period. It is probably not coincidental that both of the Slovakian cemeteries are located near the pass leading further towards Volhynia. It is not hard to conclude from the situation mentioned above that we are faced with societies which carried out intensive exchanges, mostly being the high quality flint from Volhynia as well as copper from the Carpathian Basin. Perhaps other goods which have not left traces in archaeological material were also involved[70].

What could these flint nodules be in the context of the arguments presented so far? They had to symbolize great wealth as large amounts of sought–after macrolithic blades could have been made from them. However, unlike the blades themselves, which were countable, the flint nodules were a source of material so great that it was uncountable. It seems that in times when one macrolithic blade represented great value, the deposit of flint nodules in the richest burials of the Tiszapolgár society in Slovakia played that particular role. They were probably one of the attributes of a position which might be compared to a millionaire of today.

And what did this situation look like in Volhynia, from where flint was excavated for the production of the sought–after blades? The known burials of the Eneolithic societies from Volhynia also contain fragmented blades of a similar structure as those to the cemeteries from the Carpathian Basin. However, this argument would need to be clearly defined. When comparing the inventory of both groups some clear differences are visible,

---

[68] Šiška 1964, 293–356.
[69] Lichardus–Itten 1981, 279–283.
[70] Sherratt 1982, 13–26.

which may be seen as a dialectic relationship based on the fact that small blade fragments clearly dominate in cemeteries in the Carpathian Basin, while the larger fragments–at the Volhynia cemeteries. Naturally there are many other cultural differences between these two societies which are known in specialized research literature, nevertheless in both cultures we may distinguish identical size classes of blades. The only difference is that they appear in the graves of both cultures in exactly opposite proportions (Fig 4.5; b, c). Because of this we should not ignore the impression that although the analogical metric selection of blades took place in both cultures, it was made from different points of view. The one factor which connected both societies was the fact that the fragmentation of blades was approached in an identical fashion. Statistical analyses of the blade fragment length from the cemeteries of these societies clearly show that a four–part division was followed as the division of the studied variables (the blades' length) is identical with the model division of the blade into four parts (Fig. 4.5: a)[71]. It is the most efficient way of fragmenting objects whose initial length ranged roughly between 16cm and 20cm. By acting in this fashion we may simply break the blade in half and then proceed to break the half of that blade in half. Solid tools of 4–6cm in length could still be made from these fragments, and these were common in the Eneolithic. Most of the scrapers, very common tools in the Eneolithic cultures, match this size.

The characteristic and clear manipulation of the macrolithic blade length among the Carpathian basin and Volhynia societies is a significant clue. It is also important to point out the fact that the desire for macrolithic tools, that is blades and axes, is one of the most visible elements of European Eneolithic culture[72]. In the western part of Europe we find similar behaviour, although up to now little is known about their manipulation. The best known macrolithic tools were made in Rijkholt and Spiennes, which spread up to the foot of the Alps[73]. Jan Lichardus distinguished three main production centres of this kind in Central Europe: the Carpathian region, the Baltic region and the Rhein–Mass region, which supplied tools to areas of today's France, Germany, Belgium and Switzerland[74]. Of course not everywhere did the production and circulation of macrolithic objects take on identical models, which requires us to take into account local traditions every time. Unlike in the Carpathian region, in Western and Central Europe we are faced with a

---

[71] Dzbyński 2008, 108–141.
[72] Lech 1991, 557–574.
[73] Geslin et al. 1980, 289–293.
[74] Lichardus 1981, 265–270.

ritual connected with constructing megaliths which was not present in the south. As we already know, megaliths connected society into a more coherent narrative, oriented into a community, while in the societies of the South the process of separating the dead from the settlements and establishing separate cemeteries led to great stress put on individual expression, the result being that individual items started to be fitted to particular graves. Hence, we always have to examine the manipulations of macrolithic tool length in an appropriate context. When taking into consideration the former examples it is beyond doubt that the dimensions of macrolithic objects were an important element of the cultural discourse of everywhere in that period. This is reflected inter alia in the hoarding phenomenon, which is a characteristic feature of the Eneolithic period. In Scandinavia the deposit of axes near megalithic tombs was a very common custom, which gained or lost its force depending on the period and type of rituality connected with architecture. The situation was similar in Poland, where among later farming societies (the Funnel Beaker Culture) the number of grave items grew by 7 times in comparison to the early Neolithic period [75]. The deposited tools were almost exclusively macrolithic axes and blades. Depositing macrolithic tools in those societies also seems in opposition to the graves which were humbly equipped. This situation changed in the later period connected to the Globular Amphora society.

The changing cycles of deposit activity may be a clue which should be examined in a diachronic perspective. Scandinavian research seems to suggest that the best/largest macrolithic objects were deposited in the ground or other places, while what was placed into graves most often at that time were small objects, which were worn from the long period of being exchanged. In other words, all categories of objects–both new and old–are found among grave deposits, while in graves only those with a very long "biography" were placed. Moreover, an element of archaic exchange can be seen here, in which objects gained value (a narrative one) as a result of their circulation in societies and such systems became out of date later in the north of Europe. Only in the later Eneolithic period was size connected with the abstract accumulative value connected with copper production. Such a situation can be observed in the case of the Globular Amphora societies, where large axes corresponded with the age and gender of the deceased and the localization of megalithic tombs. At that time the form of a flint axe clearly referred to the copper axe.

---

[75] Kaflińska 2006, 5–26.

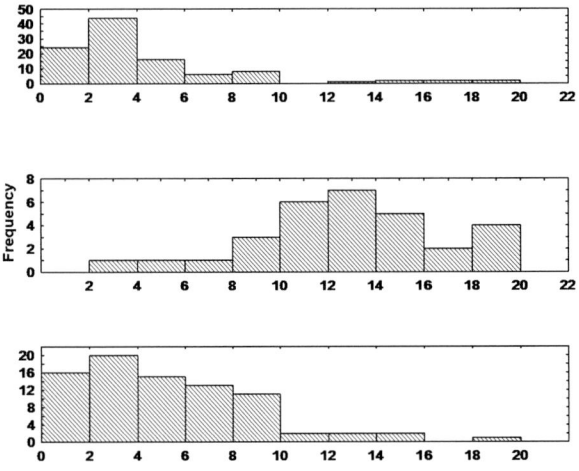

Fig. 4.5: a) The fragmentation model of the blade into four parts,
b) Sizes of blade fragments found in the graves of the Lublin–Volhynia societies,
c) Sizes of blade fragments in the Tiszapolgár graves.

Conclusions

The early Neolithic period was characterized by the close relations of the living to their ancestors, who were not isolated from households and settlements after death. This practice had a lot in common with the role and value of architecture in a time when the whole of society was almost completely connected in some way by a vision of constructing a body–cosmos order. In the Eneolithic period graves were definitively placed in separate locations outside of settlements, whereas in northern Europe houses of the deceased (megaliths), where the dead were laid, or rather built into the architecture, were being extended. It may be said that this was a kind of a trick as, although ancestors had been placed outside the area of the living, they were locked in monumental structures. In the south of Europe the development of cemeteries continued and the graves contained objects specific for each deceased, and as a result we can observe the development of social differences. But if the dead were separated from the world of the living, and some social connections cut, it became necessary to identify them by means of other cultural markers. One such marker was the macroliths which were a material transformation and a new conceptualization of the measuring stick metaphor. The area of social discourse had been opened for individual approaches, though enforcing a new means of expression, along with a new medium, keeping in mind that the medium is the message[76].

The process of moving graves away from the settlements was also linked to the development of a new symbolism of the dead, more specifically their corpses. Where the construction of monumental architecture was on the decline, the body gradually began to acquire a medial aspect connected with interceding between earthly and cosmic worlds. The orientation of the body according to cardinal directions was already emphasized in the early Eneolithic period, of which I have written about earlier in this book. Along with this, the discourse connected with drawing new boundaries between genders, which can be observed in almost all Eneolithic cemeteries, began to develop. The newly established divisions became so deep and solid that they remained the defining elements of some late Eneolithic communities in Europe (the Beaker communities). The body became the focus of that part of religious, mythological and narrative imagination which before had bonded people, house, settlement and/or megalith in a communal sphere of metaphorical expression.

Following the context presented above we should expect a proper transfer on the disposition of macroliths and other objects present in graves. Their setting could not be random as the symbolic character of the macroliths could not be left without use in relation to the body. This is distinctively seen in the burial ceremonies of the Eneolithic communities of the Carpathian Basin, where according to Lichter the larger blades were most often put next to the deceased's head while the smallest next to the feet. A unique gradation can be seen in some burial places, e.g. in grave 52 in Tiszapolgár–Basatanya the largest blade was placed next to the head, three smaller ones of the same length next to the chest while the three smallest blades, also of the same length, were put into a container at the deceased's feet[77]. It was specific for the Tibava people, as we can recall, to put large flint concretions at the feet of the dead[78]. The same was done by their neighbours – the Lublin–Volhynian people. Here the macroliths were clearly connected to the head while small tools or their fragments were placed next to the feet[79]. An identical procedure was used for small metal objects, as we will see in the next chapter. It was probably the beginning of a new symbolic and metaphoric tradition as the climax of the

---

[76] McLuhan 1964, 23.
[77] Bognar–Kutzian 1963, Plate LIX; Lichter 2001, 274–291.
[78] Lichardus–Itten 1981, 279–283; Lichter 2001, 274–291.
[79] Zakościelna 2010, 168–172.

procedures connected with the body as a reference line can be found in the late Eneolithic period of central European communities[80].

But how should we perceive the macroliths themselves? Producing these idiosyncratic tools (especially macrolithic blades) along with the number of processes connected with the organization, the production of blades and tools, their fragmentation, multiple alterations as well as the metric selection certainly found its reflection in understanding through language. Some words, concepts and elements of grammar had to evolve in the language of our Eneolithic ancestors which would help to operate the process of manipulating blade tools which were more complex than ever before. We may assume that the whole manipulation connected with macroliths was operated as part of some completely new coming to understanding process–a new vocabulary which had to include concepts which introduced designates of simple rules of divisional proportions, such as quarters, halves, three quarters, etc., into language communication.

A new element began to make its way into an archaic method of communicating, which was roughly based on exuberant storytelling which had been present since the Palaeolithic era. This element was the simple understanding of the rules of proportion that is a rational concept of measure. That is why it seems that macrolithic blades in the Eneolithic period were not only tools but also, or even mainly, special manifestations of precisely formed messages, which in some situations relieved narration through the use of a rationalized concept of number/measure. These concepts had to be used as part of the same vocabulary enabling the development of new discourse areas and creating new metaphor networks.

We can also count macrolithic tools as an individualized manifestation of the measuring stick metaphor. It seems that it did not develop from nothing in a situation when intensive discourse connected with manipulating this metaphor in an imaginative field of community which was incorporated into architecture continued for a long time. We can see the spread of a metaphoric network, linked to metrology in macrolithic tools at this time in the sphere of individual social relations. It is worth underlining, however, that a distinctive gap, a separation from cosmologic correlations to which this metaphor was assigned through monuments, was created. Macroliths became personal and had no astronomical connections in most of the activities that they were used for. Although they are still linked with the landscape and environment, the measuring stick metaphor had been included into individualized social relations even swifter without clear cosmological connotations. Because the body took on a relevant

---

[80] Kovarova 2004, 21–37; Havel 1978, 91–117.

cosmological reference (placing bodies according to astronomical directions) in consequence the connection between macroliths and the body were stronger. This is an important argument as this presents us with the moment of progressive rationalization, during which social norms became gradually detached from the primal narrative context connected with cosmological beliefs.

For the first time in European history tools gained a medial aspect in the Habermas sense[81]. We must remember that the form of the macrolithic blade resembled a thin strip of material which is more or less identical on each side. Moreover, the technique used for producing them made them straighter than blades manufactured using home methods as they were not subject to the so called bending and the top was similar to the bottom part, as the pressing techniques did not cause a distinct protuberance in that place[82]. The blade's form became constant, homogenous and geometrized. The mental image of such a product was very close to the one nowadays referred to as the bar or slate (Fig. 4.6). Macrolithic blades were a piece of a normalized valuable substance, in that way representing the same meaning level which soon was to fall upon a bar of metal. Although the bar idea was introduced to Europe hundreds of years later, first in the form of axes of which I shall write about in the next chapter, it was most probably the axe and the macrolithic blade, thanks to its metro–logization in social relations, that were a clear herald, and the first introduction, of such a path of thinking and acting.

It is no surprise that macrolithic tools are connected to the intensely developing metallurgy technique. Janusz Budziszewski performed a comparison of the two technological paths. He said that both metallurgy as well as macrolithic technology were practised on a similar socio–economical basis[83]. The person who received a copper product did not have to know how it was made as his role was to participate in the exchange. A similar process occurred with macrolithic blades and axes. On the one hand, there was the specialized production in separate settlements which generated prestigious flint goods and on the other, home production based on local traditions and resources, which focused mainly on makeshift production and altering tools by the use of primitive techniques. This was clearly separated from specialist activity, which always had a cross–regional, and often cross–cultural, spread[84].

---

[81] Habermas 2002, 481.
[82] Migal 2002, 255–266.
[83] Budziszewski 2006, 315–327.
[84] Ibidem.

Fig. 4.6 The macrolithic blade was both a requisite of prestige as well as a mental image of the metal bar.

In other words, the production and exchange of both these Eneolithic technologies followed similar socio–economical rules and shared a similar meaning. **Measure became the basic landmark element in the communication process**, which corresponded on the plane of the metaphoric network along with such features as specialized production centres, copper, joint markets, precious resource extraction from the ground, value, prestige, etc. Consequently, macrolithic products gained a medial aspect as they were a "value in their own right"[85]. They could be the subject of a peculiar thesaurization, hoards being proof of this, even when abstracting from the ceremonial character of all kinds of deposits. Acknowledging these products beyond their previous narrative contexts connected with cultural reproduction successively moved communities to a higher level of abstraction and bonded different cultural traditions.

---

[85] Budziszewski 2000, 276.

Could the development of both production branches not suggest a dialectic perspective? Let us imagine the fragmentation of macrolithic tools on one side and the accumulation of metal objects on the other. According to Chapman[86], whether through genealogy or participating in barter networks, which was underlined by certain manipulations of material culture that is the fragmentation of artefacts (mainly pottery), the dominant form of social relations in that span of time created enchainment. However, he notices the development potential of other forms of relations, which were based on spreading control over whole artefacts and correctly connects this process with the new metal technology. Metallurgy technology aided the increase of wealth accumulation, made it possible and even embodied it. Accumulation led to facilitating a new type of relationship between man and artefact which was a counterbalance to the tradition of fragmentation and enchainment. According to Chapman, as a result of this process the traditional bonds between man and object, which had its specific history and social value, began to crumble. The objects started to be formed somewhat without a person in mind but to symbolize more and more abstract values, such as wealth, which were represented by objects bereaved of the deepest social connotations[87].

Now we may include the fragmentation of macrolithic products into the above mentioned construction. If we presume that the thesis–whose starting point was the "enchainment" of social relations on the basis of fragmenting objects in the Neolithic period (although these rules first appeared in the Palaeolithic period), then the fragmentation of macroliths seems an antithesis which introduces metrological concepts, which bypassed the narrative tradition and, as a result created a gap between people participating in such a communication, into "enchainment" relations. In other words this is the same form filled with new contents–fragmentation which introduces a new quality in the place of enchainment, or rather reverses the primal tradition. Narrative values are replaced by the first numerical concepts. Metallurgy technology became the synthesis in this case and which created a completely new quality in human relations–a **measurable accumulation** *sensu stricto*, which I shall mention in the following chapters. However, a full synthesis was not reached until the Bronze Age, when metrological standards based on the weight concept were widely spread.

---

[86] Chapman 2000.
[87] Ibidem, 48.

The dialectic model is all the more interesting as it allows us to identify a double negation in this transformation, a universal law of dialectics[88]. In short: the negation of a negation is based on the fact that the old order becomes negated in the framework of its own form and then this form becomes negated. This was actually the case in the process of going from stone technology to metal technology. The fragmentation of macrolithic tools was a negation of Chapman's signum temporis–the enchainment of human relations and then itself was negated by metallurgy technology, which was the beginning of a new system of concepts and rationalized values.

We should remember that the object's use value automatically became its trade value which, like a lens, focuses on the network of interpersonal relations. The mutual relation is certified when bartering is performed according to the expectations of the other person. According to the analytical approach, the final goal of bartering (demanding an object) is not therefore to satiate our needs but to confirm the relation of the other person towards us, which is connected to some surplus or excess. The excess, which was created on the grounds of exchanging macrolithic tools, concerned their metrological information. The barter situation itself was both old and new, which makes perceiving the fragmentation of macrolithic tools in a "straight perspective" often hard. Only looking from a certain angle or even awry lets us see its new sense. The idiosyncrasy of macrolithic tools was undoubtedly new, connected to the fact that they were measured, that is, assessed in a numerical fashion (metrological–numerical). The dimension was an excess which required a new symbolic and social use[89]. That is why it was used in a new vocabulary, metaphor and in consequence more and more complex social structures than ever before in the Neolithic period. Otherwise the metrics of macrolithic tools observed in archaeological material would be grouped randomly, which they are not. It would also be impossible to define the real development of a society's complexities.

Thus, manipulations connected with macroliths have a lot more in common with the new phenomenon of accumulation, as well as anticipating it. In its early stages there could not be any true accumulation, as flint did not have the same properties as metal. Although all the knowledge we have of the production, exchange and symbolic meaning of flint and stone axes in Eneolithic Europe points to the serious inspiration of metallurgic production, stone technology could not substitute metal

---

[88] Kołakowski 2009, 57–67.
[89] Žižek 2003, 144; 1997.

technology entirely. According to Whittle, the accumulation phenomenon does not fit the Eneolithic axes as well as other flint products[90]. All tools from the Stone Age eventually crumbled and in time returned to nature. Metal production changed that.

How can we evaluate the value of metal goods? This was a problem in the beginning, which I shall elaborate on in the following part of my book.

---

[90] Whittle 1995, 254.

CHAPTER FIVE

THE METAL AGE TRANSFORMATIONS AND THE INTRODUCTION OF PORTIONS

Mysterious beads of the Cortaillod Culture

If we take into consideration what was said in the previous chapter we can assume that the measuring stick metaphor was "incorporated" into individual human relations in the Eneolithic period by the use of macrolithic tools, which were the most characteristic equipment at that time. While not negating in any way the existing interpretations of these items as functionally exclusive, prestigious tools, or the tools of other characterizations, we may offer another one. Its consequence was the implementation of the concepts of numbers/measurements into society, thus rationalizing the exchange and communication relationships in Europe[1].

In this chapter I would like to present further transformations of the measuring stick metaphor. The Eneolithic period was also a time of the intense development of metallurgy technology which required a new vocabulary to describe it. As I have previously pointed out, we should assume that before this vocabulary and new conceptualization of metal was established, people used previous descriptive methods, to depict flint technology. The copper beads which came from the north–west part of the Alps, where copper objects were manufactured in great numbers in the Eneolithic period and then spread across most of Europe, are the key to understanding the problem presented here. The first discovery was made in the 1960s. I analyzed them statistically and described them some years ago[2]. Let us recall some of the basic facts.

---

[1] Dzbyński 2008.
[2] Ibidem; Dzbynski 2008a, 36–44.

Fig. 5.1 Two copper strings found at Seeberg, Burgäschisee–Süd (after Strahm 1994).

Seeberg

An interesting find of copper beads was discovered in the Cortaillod society settlement in Seeberg, Burgäschisee–Süd [Canton of Bern in Switzerland] which were threaded on two leather strings (Fig. 5.1). Small knots tied at each ends of the strings protected the beads from accidentally falling off. All the beads were well preserved and only showed a small but diverse level of oxidization which in effect led researchers to conclude that they were not always together on both of the strings. There were 36 of them (K2) on the longer string and 18 (K1) on the shorter one. All of them had some signs of wearing in the form of wear–and–tear, scratches etc.

The archaeologists who performed the analysis of this very interesting discovery emphasized two important observations concerning these objects. Firstly, the specific number of beads on both strings reflected a simple mathematical proportion. Secondly, they clearly differ in weight in such a way that there are twice as many lighter than heavier beads. The weight of the beads threaded on the string containing 36 pieces (K2) ranged from 0.6 to 8 grams while on the string containing 18 pieces (K1) the weight dispersion ranged from 7.2 to 17.3 grams (Table 5.1). It is clear that the weight range of both strings overlap each other slightly (ca. 0.8 grams). However, as we are concerned with the period in which precise measurement systems were unknown we should perceive this as more of a clue. The arrangement of the beads on both strings was surely not connected with such a method of measuring their weight similar to the methods used today, although it is characterized by its great effectiveness.

Among other reasons, this is why the first researchers declared that the Seeberg beads are not (exclusively) ornaments[3].

Six types of beads were distinguished during the further studies of this discovery; their chemical composition had been analyzed in search of analogies which would confirm that the material used for making them probably came from the Northern Alps. Moreover, the authors of these studies stated that they do not have the same weight or shape, which would exclude the idea of them being objects of a standardized value. Such an option was simply not taken into consideration at that time.

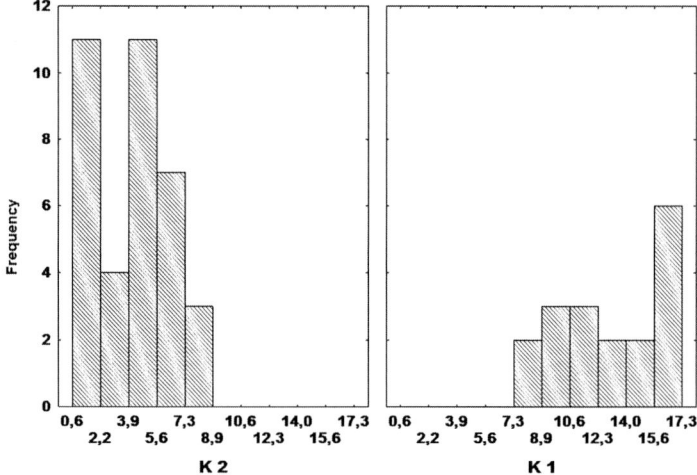

Fig. 5.2 Distribution of the weight variable for beads recovered from Seeberg, Burgäschisee–Süd. K2–longer string of 36 beads; K1–shorter string of 18 beads.

However, some important observations have been made which allow us to see these findings through a new perspective. When focusing on the method these beads were made with, it was noted that they were made from 3 or 4 metal sticks. These sticks were divided into a specific number of beads and coiled up to make them easier to transport. In this way, according to Strahm and Sanhmeister, we would be faced with an early form of an easy exchange object–a bar. Both beads as well as their original form–the stick from which they were made–were a form of the bar which was manipulated in quite a complex way, as we shall see below[4]. Some

---

[3] Sangmeister & Strahm 1974, 189–259.
[4] Ibidem.

years later it was proposed that the Seeberg beads should be treated as special purpose money[5]. However, the most important question remained unanswered: if these objects were a sort of currency then how was it used? How was it valorized and counted in the Eneolithic period?

Some years ago I undertook a more detailed statistical analysis of these findings which allowed me to discover a feature of this deposit that was unknown before. Let us have a look at some simple statistics (Fig. 5.2; Tab. 5.1). The distribution of the beads' weight from each string (K1 and K2) adopts a bimodular form, which means that two weight categories had been placed on each of them. Not all weight groups are marked in the same precise way, which is probably determined by the influence of the post–deposit processes and the fact mentioned that they show different wear–and–tear (beads differ in their "biographies"). However, the observation of these sub-distributions is not a problem.

The values of the four defined weight groups are shown in table 5.1. Although these values are approximated, we may observe that the whole variable seems to be defined by a common denominator, in the estimate 5.6 grams [group 2]. The next group is the duplication of this value [group 3], while the subsequent one is a triplication [group 4]. Group 1, whose value should be half the value of group 2, which is about 2.5–2.7 grams, is the least precise. This inaccuracy may be caused by different factors, most probably wearing processes, which are more visible on the smaller beads.

Hence, the Seeberg beads seem to present a simple metrological structure based on the manipulation of the basic values, which is about 5.6 grams. If we assume that we are faced here with a metrological structure then such an interpretation certainly seems to be logical because of the following. The oldest metrological systems were most often based on simple rules of proportion, that is doubling and halving specific units[6]. For example, if we assume that the basic unit in this system is defined by the last maximum of this variable–about 16 grams [group 4 from the table], half of this unit would fall exactly between the maximums of 2 and 3 [group 2 and group 3], which means that they would not match the observed distribution. In reference to the beads, this rule is best followed by the 5.6 gram value, it also the most often represented in the whole collection[7]. This value is quite well matched with, inter alia, the Anatolian

---

[5] Ottaway & Strahm 1975, 307–321.
[6] Hemmy 1938, 601–612; Jansen 2010, 125–129.
[7] The ideal distribution of the beads' frequency in groups 1 and 2 would of course be the following observation: group 1: 12 pieces ; group 2: 24 pieces. In such a situation all groups, as well as the whole collection, could be divided by 4.

shekel; similar units of weight were also used in the Aegean area, although hundreds of years later [8].

Tab. 5.1. Summarizing the analyses of the Seeberg beads' weight (compare Fig. 5.2)

	N	Average in grams	min	max	Standard deviation
group 1	12	1.55	0.57	2.54	0.65
group 2	22	5.69	3.49	8.05	1.23
group 3	12	9.74	7.28	12.80	1.74
group 4	8	16.05	14.74	17.32	0.90

In order to search for metrological structures I used the Broadbent method in my early analyses. This method was based on taking into consideration the normal distribution as well as examining if individual sub-distributions of the variable, tested for any existence of a metrological structure, are placed at equal intervals from each other[9]. However, the need of having appropriately numerous variables was its weakness. Results from using Broadbent's method may be confirmed at present by the use of another statistical method used to confirm the existence of metrological structures. Kendall's analysis offers such a possibility, by use of which Kendall himself disproved Alexander Thom's thesis. As we remember Thom's measurements could not offer sufficiently precise data. However, this is not the case with such findings as the copper beads.

---

[8] Petruso 1992.
[9] Broadbent 1955, 45–57; Dzbynski 2004, 29–32; Petruso 1992, 71.

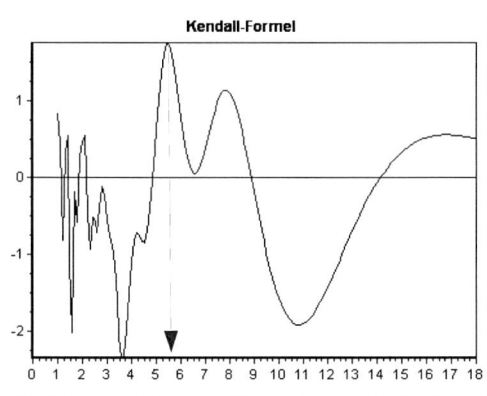

Fig. 5.3 The Kendall analysis of the copper beads from Seeberg, Burgäschisee–Süd.

Kendall's analysis is not limited in the way Broadbent's method was, although he also sets himself other goals. Rather than performing a holistic observation of the distributions, the analysis only allows the selection of the most probable value which takes part in the doubling rule in the metrological structure. This value is then defined as the sought unit of measurement. Kendall's method is commonly used by archaeologists who study early metrological systems[10]. In our case the analysis shows the value of 5.6 grams as one of the best representing the rule of the whole collection (Fig. 5.3)[11], although–as we recall–this value should still be treated with caution.

With regards to the above evidence, may we believe in the existence of a metrological bead structure, which was a specific means of communication among the Cortaillod society? Well, we can say. Yes and no. It is somewhat an illusion which results from the approach based on the assumption that there is a certain regularity enchanted in the beads–an abstract numerical rule. The right question is: what lies behind this? Is it the fact that the beads were weighed? Of course not, although everything points to the fact that they were valorized in some fashion, or more precisely a rational measuring method had been used during their

---

[10] Petruso 1992; Pare 1999, 421–514; Rasch 1987, 341–346; Rahmstorf 2010, 88–105.
[11] Kudos for the author of the software–Matthias Zimmer for the possibility of using it in this book.

production, which was in general suggested by the researchers who were the first to study this deposit.

From today's point of view, most importantly, it seems odd that four typological–weight categories had been used by members of the Cortaillod society. However, the findings made in Switzerland can easily be understood if we put them in an appropriate historical, economical and technological context. As we saw in the previous chapter, in the sphere of flint technology connected with the production of macrolithic blades we also had to deal with a specific method of dividing the macrolithic blade, most often into four parts. Specific fragments could reflect specific metrological values, such as halves, quarters, three quarters etc. The same had to happen with the Seeberg beads, which may be seen as a literal transference of these manipulations to metal. Literal, in this context, is ambiguous as it means that the verbal description of specific amounts of the material, in the case of manipulating blades as well as the beads, was the same or similar. In other words, the vocabulary used in reference to flint and copper had to be the same for some time[12]. Specific types of beads were seen not through an abstract weight measurement but a spatial one, which was more concrete, as the Eneolithic societies in Europe were still on the path to the new conceptualization of metal and the discovery of its wonderful properties. Before establishing a more abstract form of description through weight, which probably happened at the turn of the Eneolithic period, metal was evaluated similar to stone or flint, it was also processed in a similar way in the workshop[13]. The vocabulary of metallurgy in the Eneolithic period, along with the use of abstract measurement, was still *statu nascendi*. This is why we can observe four weight–type categories among the Seeberg beads.

What could the process of making such beads have looked like? The discoverers and first interpreters of these relics state that the beads were made from a copper rod which was divided into specific fragments. This rod was subject to malleable processing, the final effects being characteristic small bars, which were subsequently knotted to form a bead. Of course, we cannot exclude the possibility that the Seeberg set was created from a few rods from different time periods. Hypothetically, however, the making of such a set is possible through a certain manipulation and division of one rod, which I confirmed by writing out this mathematical riddle on a piece of paper. This process could be done as shown in the picture (Fig. 5.4). Some fragments of a divided rod had to be

---

[12] Dzbynski 2011a.
[13] Strahm 1994, 2–39.

subsequently stretched to twice or four times their original length to produce an appropriate number of beads. In other words **the production of such objects was a case of the appropriate manipulation of a metal rod by following simple rules of mathematical proportions.**

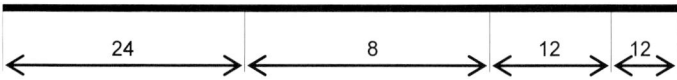

Fig. 5.4 The copper beads from Seeberg, Burgäschisee–Süd could be produced from one single rod of copper. The numbers present the amount of beads made from specific fragments of the rod.

Colmar

The Seeberg deposit was abandoned for a long time, which is why some archaeologists did not believe the hypothesis concerning its metrological structure. It is worth adding though, that these types of beads are quite common in places of the Cortaillod Culture in Switzerland, France and Germany. Unfortunately, most of them were found without any distinct context, which made a complex analysis more difficult. However, as such interesting findings exist we should include them into this discussion.

One such discovery was the site in Geröllfingen, where the beads of only two metrological categories were found. It is clearly seen when comparing histograms which concern both findings: Seeberg and Geröllfingen. The collection from Geröllfingen is much more humble and there are two sub-collections (sub-distributions). If we would treat the first sub-distribution as one then we should assume that in Geröllfingen there are two even bead groups missing (Fig. 5.5). This is another observation which clearly suggests that these beads functioned as metrological messages in social relations. Their role was the representation of the value of metal in a rational mathematical way. However, that is not all.

A breakthrough discovery was made in 2008[14]. At the Colmar site in Alsace during rescue excavation research, an Eneolithic grave with the type of copper beads with a characteristic feature of the Cortaillod society was found[15]. The localization of the objects in the grave removed all

---

[14] Kudos for Philippe Lefranc for sending me the precise information concerning the findings.
[15] Lefranc et al. 2009, 43–45.

doubts as to how the revaluation of beads was made in those times, as we will see below.

During the excavation three graves were discovered with four bodies buried inside of them. One of them caught a lot of attention as it contained the skeleton of a grown man placed in an untypical position on his front face down[16]. The grave contained only one pottery container and the beads mentioned earlier were located in a very specific way. A necklace which had 25 beads was found near the feet of the deceased. A second one was found on his waist and the third group of 4 beads was discovered under the skeleton. This localization of the three groups suggests that the beads were on strings similar to those from Seeberg and attached to the dead in some way. Perhaps his legs and waist were wrapped and the group of 4 beads on his chest were tied with an additional leather string. Researchers expressed their opinion that this burial, both with its untypical positioning, which could have been manipulated, as well as the positioning of 400 grams of copper inside it in the form of beads, has to be counted as exceptional. This must have been connected with a special event in the life of the society as well as a special person.

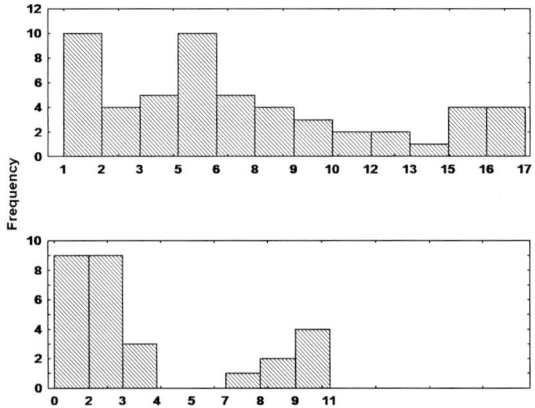

Fig. 5.5 The distribution of weight of the Seeberg (top) and Geröllfingen (bottom) beads.

In that case let us analyze the beads found next to the deceased (Fig. 5.6). The diagrams show histograms of their weight in three ranges. In the first we can see the lightest beads, with which the deceased's body had

---

[16] At the time of the publication of this book anthropological analyses as well as a full publication of this finding has thus far not taken place.

been wrapped (Collection 1). Their weight is low and ranges between 1 and 4 grams. This light weight explains the small dispersion of that collection's variable, which is why these beads seem to be equal in weight. Collection 3 is the opposite of this collection, as it contains the whole spectrum of the variable. This collection had been placed at the deceased's feet, while the collection from the chest of the deceased contained 4 beads which ranged in weight from 6 grams to over 12 grams (Collection 2).

When comparing the group with the histogram of the whole variable we can easily see the following – the deceased was wrapped with beads of the first type–weight group, the ones on his chest were the four beads from the second group while all 4 groups were placed at his feet.

Another important observation is that the distribution of weight of the Colmar beads is not the same as those from Seeberg. Does this not disprove the whole hypothesis? In short one can say definitely not. In both of these cases a similar amount of material has been used (Seeberg 382 grams, Colmar 400 grams), however in Colmar the spectrum of their weight is almost twice as large, which is indicated by the three heaviest beads all weighing about 30 grams. The heaviest beads from Seeberg weighed about 16 grams. Hence, the heaviest pieces from Colmar are twice as heavy as those from Seeberg. This means that in the case of the Colmar beads the copper stick had been manipulated differently during the making of the beads. Moreover, in the case of this collection it seems that it is less homogeneous than the former one; it contains beads which come from different bars possibly made at different time periods.

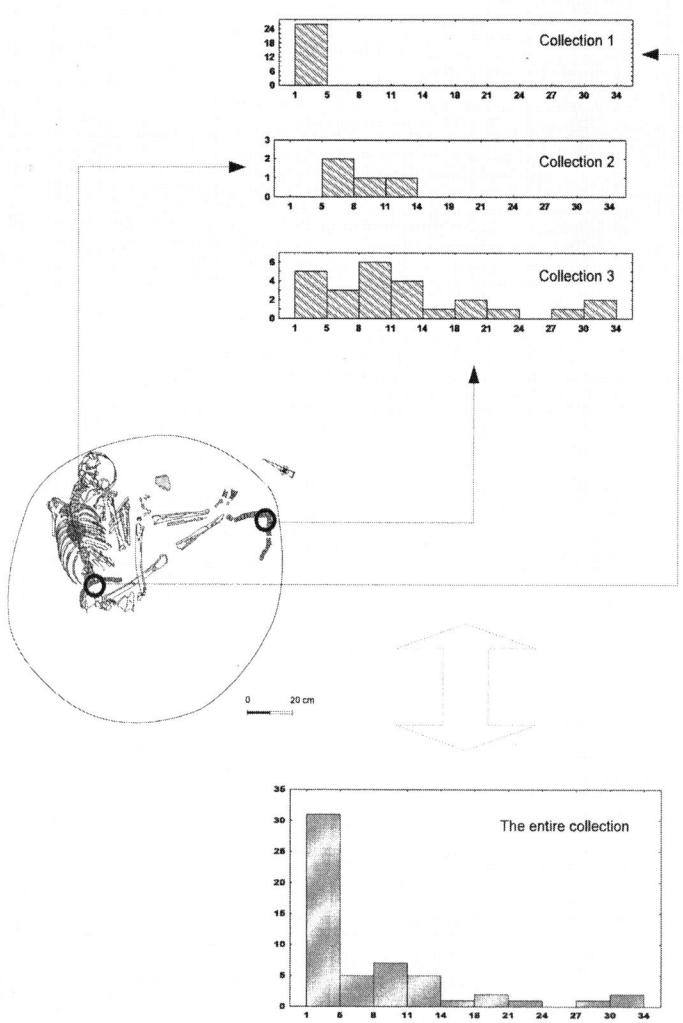

Fig. 5.6 Placement and analysis of the copper beads in Colmar. Certain values (revealing in weight categories) have been attached to particular parts of the body.

What were the beads?

In the case of the beads from the Cortaillod society we are clearly faced with another manifestation of the measuring stick concept with its transformation to objects made from metal. It bore some consequences, as it introduced a society which used metal at a higher level of abstraction in perceiving number and measure. The basic characteristic of copper is its malleability, which enabled it to be stretched and shortened, which is impossible to achieve with flint technology. Moreover, metal manipulations had to raise the question concerning the nature of measure and number in our Eneolithic ancestors' minds. The unchangeable became evanescent and so with a new means of conceptualization had to be invented for the new material. We may assume that our ancestors were aware that there was something more behind the concepts of the stick measure, something which defines that substance, something invisible, deprived of form. Developing the beads in order to measure their true numerical value did not make sense as they could be transformed by molding. Their value could not be measured in a typological way similar to how it is done by modern archaeologists. A specialist trained in using mathematical proportions had to be in charge of their production. A new substance–metal–soon began to be defined by weight better than any other material but, before it came to that, Eneolithical societies had to use a vocabulary of an older origin. In the case of beads we also can see that the measuring stick concept fulfils this role well.

Hence, as far as the Cortaillod beads are concerned, we are faced with material evidence of an ongoing discourse connected with the value of metal in Eneolithical societies. Not only that, these beads shed light on macrolithic production; more specifically they are a conceptual bond of old Neolithic concepts and vocabularies with the newly developing concepts/vocabularies of the Metal Age. Here we only observe the development of ingenuity in this direction, while the major part of manipulation has its source in the mentality of Stone Age man, not Metal Age man. The formed sticks and bars are a realization of communication processes which took place in the case of flint objects. The beads described above are something like the Rosetta stone for our divagations. They moulded the traditional approach of more archaic and less abstract measuring methods, which were made by using the measuring stick metaphor, and the new one, based on taking advantage of the malleability of metal to make a specific number of beads. **I think that in this case we may speak of the forming of a concept of portion**.

First of all, thanks to them we understand that the mathematical concepts, abstract numbers, which are well known to us today, were alien to our Eneolithic ancestors. They had been reaching some truths which seem obvious today. Secondly, the beads show us the path through which they reached those mathematical truths. At first, people learned to evaluate length. It was the basic starting concept of the measuring stick, which I described in my previous works just as a metrological concept[17]. The measuring stick idea had been confronted with the new technology of metal, which enabled such kinds of manipulations, that the earlier concrete form lost its meaning, as the longer rod, though proportionally thinner, still remained the same rod. Other features, which so far were not taken into consideration, had to be evaluated, for example weight.

But before this happened other ways of evaluating the amount of metal, which had more in common with the measuring stick idea than with the abstract understanding of weight, were used. Beads are definitely evidence of this. When we look at the charts and tables in this chapter we see that the inaccuracy of the categorization is quite large and the weight of the Seeberg beads from the two strings overlap each other. This could have been a reason for the criticism from traditional archaeologists, although this criticism is shallow, lacking the understanding of the nature of the matter. The beads were not yet categorized by weight even though these societies were probably aware of the fact that metal is defined by something more than its form, e.g. the length of a stick. The full transformation to the new system occurred only in the Bronze Age when people learned to make use of abstract weight. In the Eneolithic period weight was still a mystery, although the manipulation of beads assures us that this was a mystery which had been rationalized and defined in some way. That is why we may name this phenomenon a portion, a slightly more abstract measure than the spatial measure, not abstract, however, as the weight of the bronze bars are, which we shall discuss later on.

Different ways of measuring

While concluding the subject of copper beads in the early conceptualization of measuring it would be reasonable to accept the assumption that in this matter the spectrum of material production was probably broader than the bead–bars discovered in Cortaillod, Geröllfingen or Colmar. In reality the archaeological material shows different types of beads, among which there is one example which deserves a short description at this point. It originates from the Polgár–Csoszhalom site in the Carpathian Basin. The beads found there were

---

[17] Dzbynski 2008.

made in a completely different way in comparison to the Cortaillod ones and this is the reason why we should focus on them as the last ones.

The Polgár–Csoszhalom site is an enclosure in the type of rondel, which I wrote about in the previous chapters. It was a massive founding surrounded by four or five rings of entrenchments, in which four entrances/exits were placed according to the cardinal directions. This served as a centre for ritual activity, confirmed by the large amounts of artefacts of special meaning found there. The beads were found in a pit inside a building which also revealed such discoveries as an anthropomorphic statuette, a set of miniature cup–shaped vessels, one big vessel and 3 ceramic discs with fluted edges. The pit had a homogenous filling, pointing to the fact that the beads were deposited there only once. The whole context of this finding is unequivocally interpreted by the researchers as ritualistic–both the surrounding (the site itself) as well as the pit in the centrally located building. The localization itself, clearly connected with the contents presented in chapter 3, may explain its special purpose. The building was located in the centre of the rondel, at the intersection of the lines connecting the entrances/exits. The stage to which the findings had been attributed is characteristic for pottery that were painted red and white[18].

Let us concentrate on the copper beads (Fig. 5.7). They have a barrel shape, which is completely different from the Cortaillod beads, and they were grouped into segments. The following set of these beads were found in a cave: 63 singular pieces, 55 pieces with two beads, 24 pieces of three segments, one four–segment and one six–segment piece. The total number of beads was 259 pieces. They were probably a part of a long necklace on which 20 bone ferrules were threaded. As a whole it is an impressive amount of copper used at one time.

Let us now move on to the strange way of grouping them into segments. Despite the notable corrosion, researchers do not have any doubt as to their segmentation. Even today it is clearly visible. Therefore, the beads were composed purposefully by the maker so they would present the following sequence of values: 1, 2, 3, 4, and 6.

At this point we should consider another possibility of early metal conceptualization which we only discussed in the context of the Near Eastern notation systems presented in chapter 2. We would be faced here with accenting the counting portions of metal in a one to one correspondence and a representation of a collection of objects. Individual beads could have been separated from the rest and put back again into one

---

[18] Raczky et. al. 1996, 17–30.

sequence. However, the connection of these beads into one segment (e.g. four pieces) would still be conceptually connected with the stick metaphor. Hence, based on the example of the Polgár–Csoszhalom deposit, we can see a close relationship between these two basic methods of the mathematical valorization of metal in the Eneolithic period.

Fig. 5.7 Examples of beads from Polgár–Csoszhalom.

Measuring vessels in the Eneolithic?

If we were to search through Neolithic/Eneolithic artefacts, would we still find materials and tools which could have shaped human minds towards a greater mathematical abstraction? Pottery vessels are an obvious candidate for this role. Apart from transmitting narrative content by use of their form and ornamentation, almost all the vessels have a certain capacity which, similarly to weight, at some point had to be a mystery. Were they also the subject of transforming the measuring stick metaphor?

From the analysis of vessels from the early farming societies, pottery items from the site in Bylany (in Bohemia) appear to suggest that the vessels of that time had to be categorized in a quality–size way and rather did not have any numerical connotations. These vessels may be grouped accordingly into small, medium and large sizes[19]. I think in general, that the conceptualization of the Neolithic vessels' size was based mostly on the quality rule. The original "measure" of the vessel was equally its form and its size, which could not be, or was not attempted to be, characterized numerically. In fact it could have been redundant, as vessels in the Neolithic and early Eneolithic era were rarely involved in far–distance exchange, during which establishing metrological messages was essential. The fragility of pottery undoubtedly became a barrier in this process. The abundance of types of Eneolithic vessels seems to confirm these assumptions. At first, in the early Neolithic period only some basic types existed and the spectrum size was small. We must treat this as the initial period. It was only in the Eneolithic period that the development of the types of vessels and a significant extension of their spectrum size occurred. This period, as we know, ended in a far–reaching unification in

---

[19] Pavlu 2000.

the form of the dominance of two types of vessels: the beaker and the amphora. It seems that we may treat this period as a clue for our research.

Beaker societies

Beaker societies present as many similarities as differences. In fact, apart from the common use of beakers we are faced with such an amount of differences that we may analyze their relations in categories of dialectic opposites[20]. In the case of the Bell Beaker societies we are dealing with the *Begleitkeramik* phenomenon, which often constitutes most of the pottery items[21]. Meanwhile, among Corded Ware ceramics societies it was the beaker and amphorae which constituted 80% of the vessels in graves. Corded Ware beakers and amphorae, as well as Bell Beakers, are known to us almost exclusively from these cultures' graves, apart from the Alpine regions, where we are confronted with considerable evidence of sedentary settlements of the Corded Ware Culture[22].

Early studies concerning the Corded Ware ceramics culture in the Czech Republic (Bohemia) turned my attention towards the specific behaviour of the variable size of vessels in the context of the burial ritual[23]. To achieve this I used the reconstructed size of the vessels and compared them to geometric blocks which were closest to them in shape. The aim of this process was to acquire appropriately large data samples to be able to establish a wide spectrum of statistical analyses. As "corded ware" amphorae are very round a roughly looking sphere–shaped block had been used in order to assess their size, while in the case of beakers I used the approximate cylinder block.

The results of these early studies were highly suggestive. Both the amphorae and beakers, in the context of the noted traditional features of the burial ritual, such as age and gender of the deceased, the presence of other characteristic objects in the graves, presented significant differences as far as size is concerned. For example, the comparison of the deceased's age with the capacity of the amphorae present in the graves resulted in a characteristic plot curve which shows that the smallest vessels were given to children and elders, while the largest were given to adults. Moreover, on the basis of the beaker capacity, which was always deposited in male graves, a relatively distinct line could be drawn in connection to the capacity between a child and an adult. Also the presence of other objects,

---

[20] Strahm 1992, 163–177.
[21] Strahm 2004, 101–126; Besse 2004, 127–148.
[22] Strahm 1992, 163–177.
[23] Dzbynski 2001.

such as stone axes and mace heads had a rational influence on the size of the vessels placed in the graves[24]. The manipulation of the vessels' size in the Corded Ware culture in Czech area seemed to be a sign of mysterious and very complex behaviours, which, in a significant way, were involved with all other elements of the material culture in graves. In the context of these results I have proposed a thesis according to which individualized socio–economical relations of the members of the Czech Corded Ware society had their reflection in the manipulation of ceramic vessel size[25].

The results of the analyses of the Corded Ware beakers in the Czech Republic were so promising that I decided to study this case further. By broadening the query I found a very interesting and rich collection of Corded Ware ceramics from the area of present day Central Germany, where once this society had settled intensively. The flaw of these materials was its lack of reference to the age and gender of the deceased. The only thing I could have done in such a situation was to measure these vessels precisely with the use of a granular material and analyze the variable.

The analysis of histograms suggested that among the Corded Ware society in Central Germany a rule of doubling the size during the production of vessels was in use, the most precise and credible results came from analyzing the beakers. I was even able to set the basic unit of capacity for the beakers as approximately 0.17 litre. It was the smallest unit, which could have been doubled and added so as to describe the whole variable by using this unit (Fig. 5.8). However, some time later I realized that I had made the same mistake as Alexander Thom in my study of megalithic constructions in Great Britain. It will be safer to assume that the basic unit in the case of cord ware beakers would be something close to a portion, which at the present would be a rough equivalent of a medium cup with a capacity of approximately 0.2 litre.

Kendall's analysis, which I present for the first time (Fig. 5.9), supports this thesis as in the earlier studies I made use of Broadbent's method. At present, Kendall's method allows us to authenticate the results of earlier observations, which could seem incomplete to archaeologists set on complete static proof. Apparently the figure of Kendall's analysis shows that the greatest possibility for a unit of capacity is 0.2 litre. It would correspond quite well with previous observations. The metrological system of the Corded Ware Culture was therefore built on a simple rule of operating a basic unit of the system–a portion.

---

[24] Dzbyński 2010, 179–183.
[25] Ibidem.

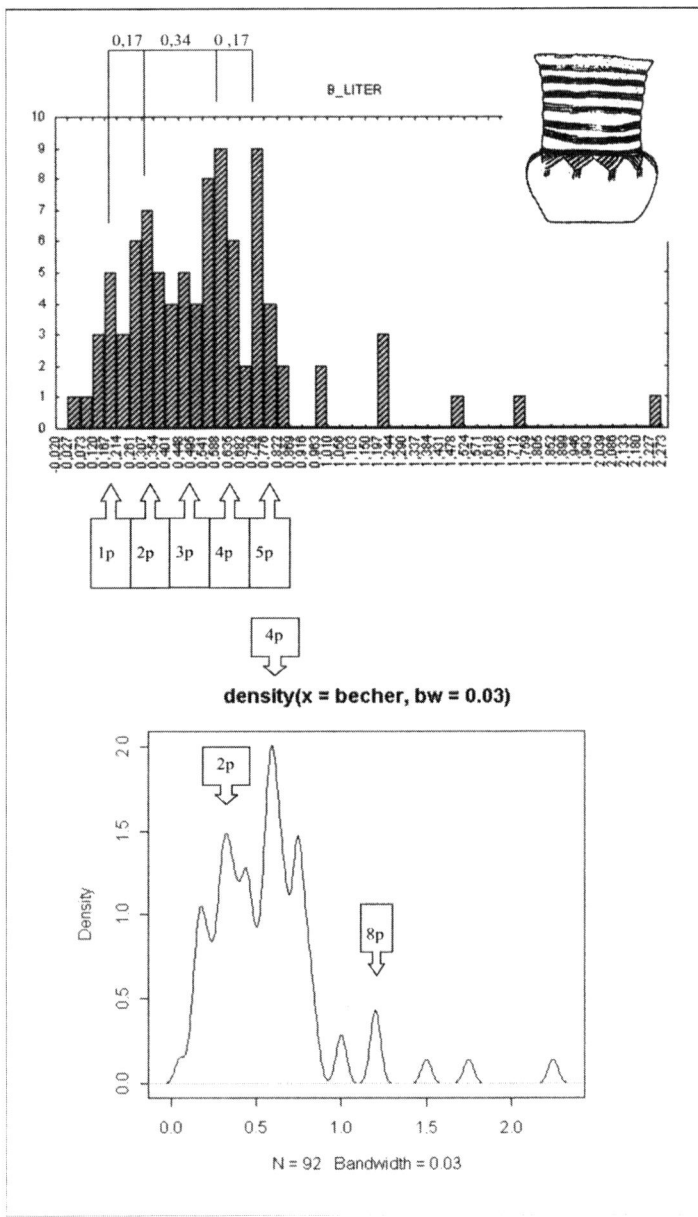

Fig. 5.8 Metrological analysis of the Corded Ware beakers from Central Germany (after Dzbynski 2004).

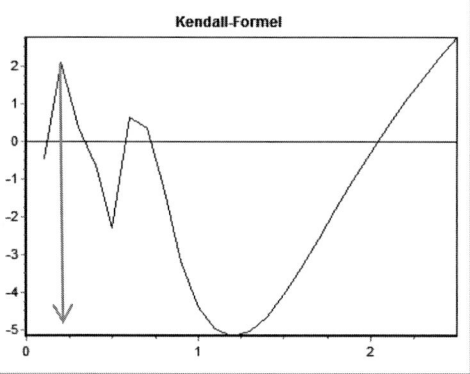

Fig. 5.9 Kendall analysis of the Corded Ware beakers from Central Germany.

And what were these portions? In my earlier research I leaned towards the idea that the portion was an anthropological value based on operating fistfuls. In order to examine this hypothesis closer I performed an experiment with a group of students some time ago the purpose being to measure granular material several times by the use of fistfuls. The students, divided according to sex, measured one portion (1 fistful) and two portions (2 fistfuls) and recorded the results of their measurements.

These experiments showed that the portion of Corded Ware beakers fitted almost perfectly in the hands of today's women (Fig. 5.10). The observation that men behaved in a more unpredictable fashion, while women tried to subject themselves to the rule of measuring a set portion, is fully understandable and coherent with the theory of psychological evolution. When taking into consideration the fact that prehistoric people were shorter than those of today we may assume that a modern woman's hand would probably match that of prehistoric men. However, can we be sure that it is the male portions which are reflected in Corded Ware beakers in Central Germany?

The concept of portions in the context of this discussion may also be put into a new perspective. It is commonly known that pottery vessels in the periods which preceded the potter's wheel were often made by use of the roller technique. Small vessels could have been made by kneading a clod of clay, while very big ones were coated with more clay in a prepared cast. For example, among the Eneolithic societies of Eastern Europe a bag filled with sand was probably used[26]. It seemed to me earlier that a small bag could have also been used by the Corded Ware societies in Central

---

[26] Cvek 2004, 279.

Europe, after filling it with the desired amount of portions – fistfuls, although it was not the only possibility.

Another possibility was making a vessel by the use of a roller, which could have effected another transformation of the measuring stick with success as the appropriate length of a roller of clay could have been efficiently reshaped to acquire the desired capacity of the vessel. When I decided to perform an experiment during which students kneaded vessels from measured rolls of modelling clay, it appeared that the larger the portions the less precisely they were made. When overlaying the results of this experiment onto prehistoric data a similar occurrence could be expected. Hence, there is a possibility that the portions 4p and 5p marked on the chart (Fig. 5.8) could in reality be the same portion. This matter cannot yet be settled. Despite this it does not seem probable that the procedure of setting the sizes of vessels in the Eneolithic Corded Ware societies had no influence on forming a mathematical sense of assessment, calculation and the ability to use proportions. If we acknowledge that the metrological structures in Corded Ware societies are in fact a reflection of conscious actions then we must assume that these actions stood behind the first mathematical concepts.

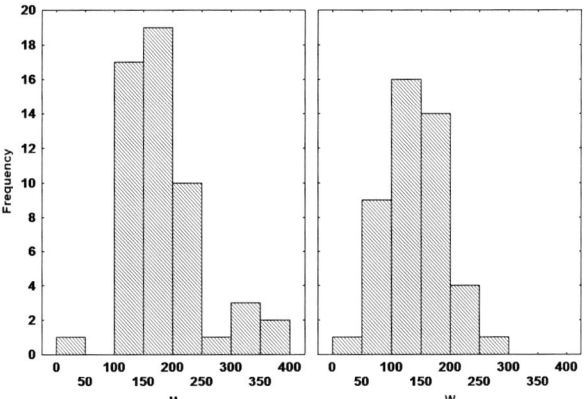

Fig. 5.10 Results of the experiment based on measuring fistfuls of grain (1 portion: M–men, W–women).

Other examples of measuring vessels in the Eneolithic Period?

The metrological structures of the Corded Ware ceramics were for some time a separated example of a rationalized form of numerical communication in the European Eneolithic period. I would not be far from exaggerating if I say that for a long time it was a mystery which was impossible to explain in any coherent way. That is why for some time I did not focus on this problem. However, on the other hand, the existence of such a type of behaviour at the end of the Neolithic period seemed logical, especially as in the Bronze Age the amount of evidence of using metrological systems greatly increased[27]. Because archaeological sources clearly show a continuation of the earlier cultural processes in the Bronze Age, the continuity of cognitive development had to be taken into consideration as well. In that case there had to be intermediate stages in reaching mathematical truths, which today may be identified inter alia with metrological structures in the Corded Ware Culture.

In time the amount of evidence which proved that the metrological structures of Corded Ware vessels are not an isolated case, increased. Regardless of the fact of finding copper beads, which presented a clear metrological structure[28], inside a Corded Ware amphora from Kelsterbach, the materials from the Bell Beaker Culture has recently supplied us with evidence for the existence of a rationalized metrological system. Vincent Georges presented an analysis of the capacity of Bell Beakers in north-western France, which come from several sites associated with megalithic constructions. When analyzing the variables he came to the conclusion that several characteristic leaps occurred which present a metrological structure in the same way as Corded Ware materials do. According to Georges these types of behaviour may be interpreted as an element of farming activity. He turns our attention to the fact that farming activity requires the development of rationalized planning techniques just as the widely understood crop administration does. He even proposed perceiving the characteristic ornamentation of Bell Beakers as parallel symbols which eased the measuring and distributing of different types of grains[29].

Cases' studies are an older example of research concerning Bell Beakers[30]. He reached similar conclusions to those presented by me

---

[27] Lenerz De–Wilde 1995, 229–327; 2002, 1–23; Neugebauer 2002, 25–40; Peroni 1998, 217–224.
[28] Witter 1941, 70– 80; Behn 1938, 77–78.
[29] Georges 2006, 609–613.
[30] Case 1995, 55–67.

together with Roland Wiermann in the past in reference to Corded Ware at the cemetery in Vikletice[31]. Based on the grave materials he showed that the capacity of Bell Beakers is clearly related to the gender and age of the deceased. He analyzed 301 capacities of beakers and concluded that vessels with a capacity above 2.5 litres more often than not came from settlements and only from male graves, while small or medium vessels were placed into children's graves.

When taking into consideration the above mentioned evidence we may conclude that in the Neolithic period there are no signs of manipulating the sizes of vessels in the context of graves. However, the end of the Eneolithic period is the complete opposite. The differences in the capacity of vessels placed into the graves of men and women at that time become statistically crucial. This includes both amphorae, beakers, which I studied in my earlier works, beakers studied by Case, Georges, as well as other vessels, which is clearly shown in Petr Krištuf's studies[32]. Hence, it is beyond doubt that in some societies in the Eneolithic period, vessel size in the social context were consciously and commonly manipulated, which was not present in Neolithic times, although in this period we can find large storage vessels in settlements. However, today I would surely not seek metrological structures in all kinds of Eneolithic vessels, although without more precise research it is difficult to say anything more. It seems to me that we should rather see this matter through the perspective of the culture specifics of a given society, which we are able to define on the basis of the whole material culture and other sources. What we can say more precisely, is that the introduction of this phenomenon among beaker societies–heirs of the earlier Eneolithic societies among which the first numbers/measures were mixed into social relations in the framework of individualized communication, seems coherent and logical.

Measuring beaker–the drinking beaker

Metrological structures enchanted in the beakers of late–Eneolithic societies do not allow us to decide whether vessels were measured by fistfuls or by the use of a roller. Of course, using one method did not exclude the knowledge of the other. Before the invention of the potter's wheel vessels were made from rollers. This was a process of transforming length into capacity–a transformation of which people were aware of in that period. However, the use of fistfuls suggests a tighter relation with the

---

[31] Dzbynski & Wiermann 2002, 205–226.
[32] Krištuf 2005, 69–118.

body which was the basic source of mathematical perception, as I wrote in chapter 1. Hence, both methods may be treated complementarily. All past analysis clearly suggest that the aim was to achieve an adequate capacity of a vessel which fitted a given sepulchral situation while the measuring activities were elements of a ritual connected with burial. However, if we assume that the starting point among Neolithic societies was the incorporation of the measuring stick metaphor in different social contexts, then among late–Eneolithic societies we can observe further consequences of this, which led to another qualitative transformation. The length measure was remade into a capacity measure, during which the process of more abstract mathematical concepts were probably established. Pre–metrical and pre–numerical evaluation, examples being the earlier Neolithic vessels, became reformed in the Eneolithic period into a numerical value.

Nonetheless, we should not forget the origin of beakers. Andrew Sherratt put forward a hypothesis, according to which the greater popularity of the beaker or cup in the Eneolithic societies of Europe was based on a certain cultural movement from the Near East, whose main elements were symposia–carousals in the company of a new social elite[33]. Alcohol, which was probably made from grapevines, was a relatively late invention in prehistory, and in Europe people used beakers probably to drink mead or some sort of kumis. This second option seemed particularly close to Sherratt as the phenomenon of communal male drinking was, according to him, connected with the development of pastoral life and the shaping of the male warrior identity. As Hodder wrote, this caused "an alternative discourse concerning power", which influenced the opposition towards the domus metaphor more and more. In other words, the typically Neolithic means of communication based on manipulating architecture met with alternative ways of transmitting authoritative relations, which were more dynamic, while "waging wars, hunting and drinking became the idiom of social discourse"[34]. Warrior–shepherds introduced the custom in Europe, which is a basic element in modern culture. Hodder also assumed that this had a symbolic meaning of taking women out from the sphere of the wild, which they were part of at the start of the Neolithic period, and into the role of the keeper of the fireside or a man's property.

The history of alcohol may also be seen from a wider perspective. The Sumerian Epic of Gilgamesh is the oldest mythological text which offers an image of the worldview of people who were the subject of civilizational

---

[33] Sherratt 1986, 81–106.
[34] Hodder 1990, 175.

processes over a relatively short time. In this epic we are faced with an image of a "domesticated" man's worldview which is distanced from nature. Gilgamesh represents civilization and culture while his main rival– Enkidu, is the personification of the wild and untamed nature. When they meet for the first time they begin to fight for supremacy. However, before this occurred Enkidu had been somewhat "humanized" by the use of food and beer, which resulted in giving him a human form. Hence, in the case of the wild Enkidu, alcohol played a culture–creating and civilizing role.

It is beyond doubt that the symposia, of which Sherratt writes in his works, were not simple social meetings whose goal was to reach a characteristic and pleasant mood. It was also a specific way of experience, participation and identification connected with myth and ritual, metaphorically connected with other significant inventions of the so called second Neolithic revolution: a broader use of animals, metallurgy, a different approach towards women, etc. Consequently the civilizing role of the social consumption of alcohol is unrelated to intoxication with various psychoactive substances which define the worldview and the spiritual life of the Palaeolithic hunters[35]. Alcohol became an important symbol of culture and civilization.

Sherratt proposed to see beakers as drinking vessels, attributes of a person who enjoyed alcohol and who travelled on horseback. However, the drastic impoverishment of the typological spectrum in the "beaker" societies does not match the information concerning their farming character. Hence, they were not strictly nomadic or pastoral societies concentrated on consuming "liquid bread" from a beaker, in which signs of grain were more often found than signs of alcohol. Apparently the beakers were used for various tasks in a household, as Georges suggested in reference to the Bell Beakers[36]. According to his research, beakers were more likely universal vessels used for all purposes. They were also special measuring tools, which are confirmed by both Georges' research and mine. That is why the phenomenon of typological unification which occurred at the end of the Eneolithic era may also be, in my opinion, seen as an expression of new needs and behaviours in those societies, more precisely as an expression of the desire to control a part of the growing number of goods by the use of measure and number, which does not exclude Near Eastern influences. This is quite the opposite! However, if we combine Sherratt's hypothesis with the results of recent analyses we must admit that vessels in the beaker societies were also a mathematical metaphor. A

---

[35] Wierciński 2004.
[36] Georges 2006.

beaker carried within it a sublime symbolic–mathematical message; it was both a manifestation of the civilizational role of communal drinking as well as the civilizational role of control and measuring.

In regards to the Beaker societies one more important note must be made. Their characteristic outline encompasses the standardization of grave items that belonged to both men and women. That is why the objects placed inside the graves often do not make the difference between genders any easier. Both men and women had a personal set of objects placed into their graves according to the cultural gender. This situation resembles in some way a lack of social differentiation which may be observed in simple hunter societies and is in fact interpreted by some archaeologists as an image of returning to more egalitarian relations among beaker societies[37]. However, it does not seem completely true. Just as the hunter, who had to be equipped with a bow and flint arrowheads, so did the Eneolithic man receive a beaker, a knife (flint or metal), sometimes a flint lighter and an axe or stone hand–axe for his final voyage[38]. In this fashion many men, who were equipped in a very similar way, create an impression of social egalitarianism, which turns out to be false if we take into consideration the capacity analyses performed on vessels present in the graves. Then it would seem that e.g. mace heads were less valorized than stone axes and their bond creates a hierarchic relation, not visible earlier in the archaeological material if we use classical analyses[39]. All in all it seems that the capacity of vessels had a basic definitive meaning, which means that social structures at the end of the Eneolithic period were already immersed in the idea of measuring and in that act they expressed social differences. **In short, beaker societies anticipated the forthcoming Bronze Age which was focused on the weighing of metal**[40].

A fitting illustration of the above is the cemetery in Vikletice, the largest burial place of the Corded Ware Culture discovered so far. There was a designated group of graves located in its centre, in which almost all the symbolic material, which was gathered from the whole cemetery, was deposited. They were stone axes and mace heads, stone cleavers as well as small copper artefacts. As it is easy to guess, in this group there were also the largest vessels, several sizes larger than the average size of vessels present at the cemetery. More than ten vessels were deposited in one of the graves, while no more than two vessels were present in the rest of them. Hence, we are undoubtedly faced here with a clear display of inequality,

---

[37] Müller 2005, 24.
[38] Budziszewski & Tunia 2000, 101–135.
[39] Dzbyński 2010.
[40] Pare 1999.

which is further emphasized by the separated localization of the burials. This group can remind us of some sort of an elite retinue and brings to mind a connection with the later social organizational models of the Bronze Age[41]. In consequence, I would be willing to acknowledge that in the Europe of that time a fundamental imperative of the Bronze Age had been formed, which was based on the final integration of all countable/measurable goods, especially metal. Without the ability to count and measure the control and accumulation of goods would have been impossible.

The European way–from stick to bar

Janusz Budziszewski points out that the establishment of the macrolithic industry is linked to the development of early metallurgy, as the production, distribution and organization of macroliths was the same as the one used for copper products[42]. After all, many societies which manufactured and used copper were also interested in producing macrolithic tools from good materials. The reason for this might be that, as I have suggested earlier, **macroliths were a sort of *alter ego* for metal in a time when both technologies were continued as part of the same complex of words, metaphors and concepts.** The basic activity during the distribution and exchange of these idiosyncratic products was their fragmentation, the *signum temporis* of the Eneolithic period which, according to the enchainment theory, was supposed to be a ceremonial means of communication, which was covered in detail in chapter 4.

Nevertheless, the fragmentation of copper axes remains a certain mystery, especially in the above mentioned aspect, and that is why we will now deal with this issue. Most signs of fragmentation were discovered in the region of the Alps, where they were manufactured and exchanged intensely (Fig. 5.11). It was also the region where the first axe bars were formed near the end of the Eneolithic period. Bearing in mind that metal was not fragmented in the Balkans and that full technological possibilities were not yet discovered[43], should we not connect these facts into one hypothesis?

First, let us analyze the weight of Eneolithic axes from the Alps region. Although associating axes with a specific period in time raises some difficulties, with the large population at our disposal we may see the

---

[41] Dzbyński 2010.
[42] Budziszewkski 2006.
[43] Schmitz 2004, 560.

general trends. In the diagrams we see that early Eneolithic axes reached a very wide spectrum of weights (Fig. 5.13: Time I), which is congruent with the observations of most researchers. Very heavy axes, which weighed almost a kilogram and which were not produced later on, constitute the major part of that diagram. However, smaller examples may also be found, which means that we are not faced with a homogeneous population. The lighter axes shown in this diagram are, in my opinion, a testament of the fragmentation and processing initiated at that time, which intensifies in the subsequent period.

Later on (Time II), a visible decline of the weight of the axes, which would be in agreement with the hypothesis concerning the rapidly depleting deposits of the available material. The characteristic trend of that time is the production of small copper objects and small or thin axes[44]. However, the depleting deposits hypothesis has its counter–disputants, who point to the phenomenon of the wider spread of copper among Eneolithic societies. More attention should be given to this hypothesis. Perhaps metal did not suffer from wearing, hence becoming more accessible for Eneolithic societies finding their reflection in making smaller and smaller forms. Moreover, one more factor should be taken into consideration, namely the specifics of metal and the emotions connected with them, which in effect culminated in trials at developing precise methods of measurement. Before it came to this the participants of exchange were aware of the basic fact that even a small piece of metal was valuable and should not be thrown away like flint flakes. That is why during the distribution and exchange cycle some behaviour could have occurred which was aimed at accumulating (from today's point of view–profit), and this may be best described with the use of the games theory[45]. Hence, it was not so much the declining amount of metal as rather other exchange strategies that had been used which took into account the growing demand for metal, or more probably its prestige–making role. This would explain the making of smaller and smaller objects.

---

[44] Ibidem, 564.
[45] Straffin 1993.

The Metal Age Transformations and the Introduction of Portions 161

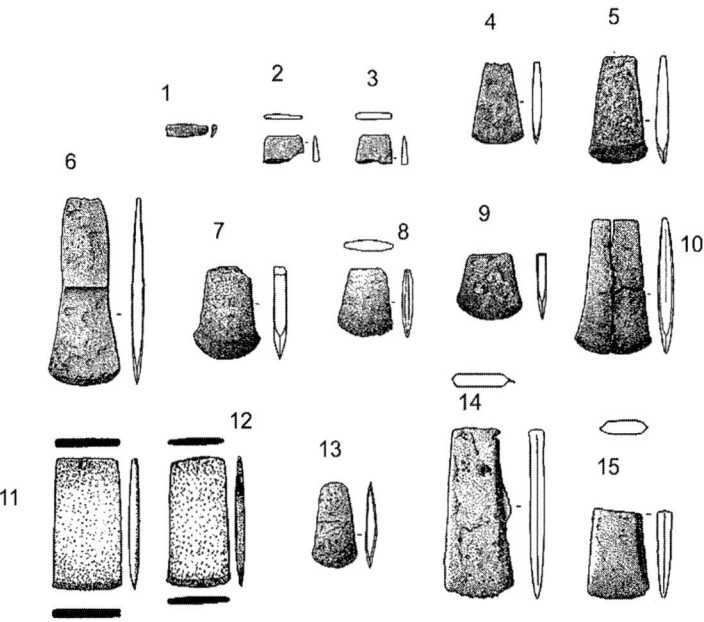

Fig. 5.11 Examples of copper axe fragmentation in Central Europe (1–10, 13: Eneolithic axes of different types, 11–12: Altheim type axes, 14–15: axe–bars).

The above mentioned model essentially supports Staaf's analyses[46]. It presents a schemata of the development of the acceptance of the axe in Europe, in which one pattern comes up persistently. The introduction of metal in the main regions of manufacturing copper in Europe was subject to growth and spread. In the first stage it affected the Balkan–Carpathian region and subsequently–central and Western Europe. What Staaf basically suggests is that by the end of the Eneolithic period there was a forming of certain norms of perception and specific activity towards metal, which he called "a general common understanding of metallurgy", something close to forming a new "mind" in the cultural discourse. In my opinion the source of this new "mind" was the innovation cycle which was based on a more rational way of perceiving the measurement of metal and an adequate description of its value, which archaeological confirmation are

---

[46] Staaf 1996, 139–152.

inter alia with the first axe–bars, probably coming from the Bell Beaker environments [47], as well as the beads, which were measured in a completely different way than those I described above[48]. It seems logical that, at the end of the Eneolithic period, the copper axe also had the value of a more meticulously estimated portion of metal.

Let us return to the diagrams. Regardless of the interpretation used, there are clearly fewer axes and they become much lighter, which is visible in the observations on the left side of the chart (Time II). The best examples of that period are the Altheim–type axes, which more resemble thin slates of metal than axes. Because of this they were easier to break and exchange, especially as the addition of arsenic resulted in a greater crumbliness.

In the third diagram (Time III), which is roughly identified with the final period of the Eneolithic period, the situation changed: the weight of axes had been reduced. Most of the axes of that kind weighed from 100 grams to 200 grams. The important fact is that the first axe bars also originate from this period[49]. It is possible that the fact that this value culminates in the weight range, which in the Bronze Age became a basic metrological standard by taking on the form of ring bars, was not coincidental.

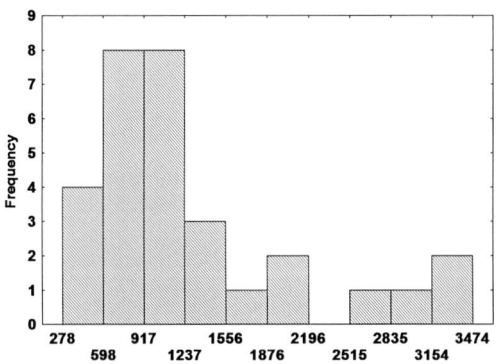

Fig. 5.12 The distribution of the weight of fragmented Zabitz–type axes.

Let us compare an additional set of data, which are the so called Zabitz–type axes. In this case we are faced with axes which were surely

---

[47] Mayer 1977.
[48] Witter 1941; Behn 1938; Dzbynski 2008a.
[49] Mayer 1977.

broken in half in order to receive half of the axe's value. Researchers accept the interpretation of them being a store of value, although the search for a common metrological unit of weight is a misunderstanding in this case[50]. This type of axe was suitable for being divided and was probably not weighed. If we put the distribution of their weight onto a chart we shall see a characteristic grouping around three values (Fig. 5.12), which point to the existence of a characteristic division into halves and a further division of the resultant halves into halves again, just like in the case of macrolithic blades (Fig. 4.5). Perhaps a similar procedure had been used in regards to most of the copper axes, which is illustrated by those of them which bear signs of attempts of dividing into two or four parts. Klassen even observed signs of sawing on some axes[51]. Assuming that the fragmentation of axes was a temporary action, after which it had been re–shaped, the number of these signs is quite large (Fig. 5.11).

However, in the end the metrological standard of the Bronze Age at first ranged between about 180 grams and 200 grams, not 3,000, 1,500 or 750, which is the weight of the Zabitz axes, along with their respective divisions. Again, the Zabitz axes only show the initial, archaic means of treating metal objects for exchange purposes. They are an exception (they are very heavy), as it was the simple flat axes which were the most common object of exchange and around which the discourse connected with the valorization of metal in the Eneolithic period occurred. It should be agreed upon that the result of this process is forming the Bronze Age bar.

---

[50] Kibbert 1980, 48; Lenerz De–Wilde 1995; 2002.
[51] Klassen 2001, 280.

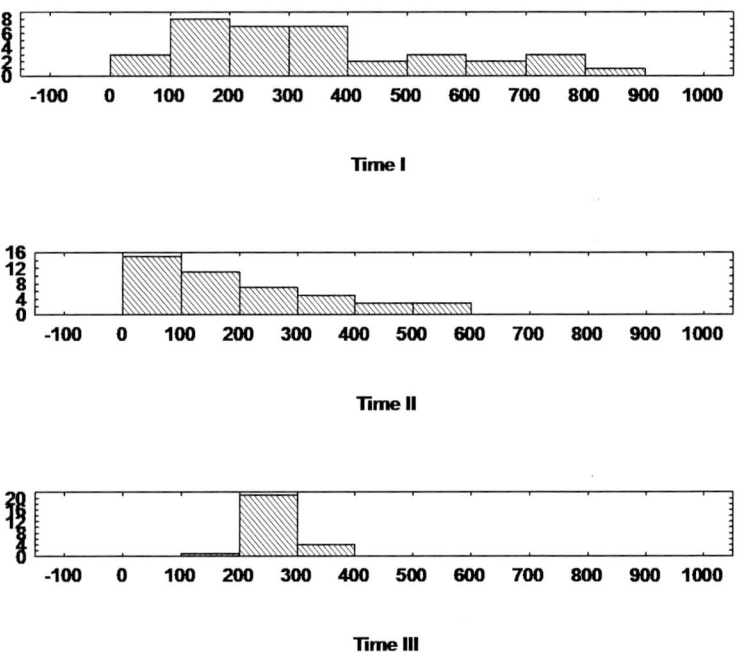

Fig. 5.13 The distribution of the weight of the copper axes in Central Europe in three chronological periods.

Conclusions

Due to the above, the course of the transformations described in this chapter, connected with the increasingly rational perception of the value of metal in the Eneolithic period, may be presented as follows. The greater individualization and internalization of the measuring stick metaphor in the Eneolithic, which took place during technical development, found its vent in the form of complex fragmentation processes, which should also be understood as a discourse on the value of metal and flint objects. Before the development of the abstract concept of weight, which is the most adequate description of metal in social relations, more specific assessment mechanisms were in use which we see as marks of the fragmentation of copper objects as well as the forming of copper into bars according to the length measure, as in the case of the Cortaillod beads and others presented in this chapter. I think that in places where copper was produced for a longer period of time the process of fragmentation quite quickly gained a similar meaning as macrolithic blades. But the effects which followed had significant cognitive and social consequences. A copper axe is more universal just not as a work tool, since almost anything can be produced: it can be re–forged into a piece of jewellery or into two smaller axes. Even a small fragment separated from an axe became very valuable and could be kept and used for probably some new item. The same could not be said about flint materials. That is why even the fragmentation and exchange of flint tools, e.g. macrolithic blades, blocked some paths of progress and was always limited in some way, while a small piece of flint could only be thrown away. The process of dividing copper axes and (more importantly!) chopping off pieces of metal should therefore be interpreted not only from the ritual–ceremonial perspective.

The properties of metal opened a new layer of underlining innovative forms of fragmentation. It was necessary to operate more abstract concepts and take on a less concrete way of thinking. This abstraction, if referred to describing axes, was followed by an increase of abstraction in social relations, as copper objects were not only exchanged but also, during that process, intensely discussed about, and a few–hundred years long discourse among copper manufacturing societies had a real influence on them. As traditional forms of description could not be enough, sooner or later, new, more abstract concepts of measurement in reference to metal exchange in the form of axes began to develop. In other words, weight, which became the basis for evaluating metal in the Bronze Age, was starting to be recognized.

At the same time, the examples and interpretations mentioned above make us aware of the fact that the process of reaching some truths, which are obvious from our perspective, took place in a time and space of which we still know little about. We may surely assume, however, that mathematics did not appear spontaneously in the heads of our ancestors and it was not introduced to them from an outside source, but was a long–lasting process, which is active to this day. At this point we have discussed only a part of this process, the very early part. Can we trust this? In my opinion yes we can. In reality the evidence presented confirms the generally accepted hypothesis that the process of forming mathematical ideas went from the concrete to the abstract. As to Europe this was also a process of transforming the measuring stick metaphor into an abstract number which belonged to a new vocabulary, describing the metal's weight, whereas, according to Renfrew, weight is a material–symbolic fact. It does not develop as an embodiment or materialization of earlier mental concepts but through the development of the concept–construct itself in connection to the experience of the material world[52]. This process took place on a human communication level in prehistory.

---

[52] Renfrew 2004, 26.

CHAPTER SIX

SUMMARIES: MATHEMATICS AND A MATERIAL CULTURE

In chapter one I presented the thesis, that the fundamental factor for the development of the mathematical understanding of reality was the development of material culture. In the chapters which followed, I tried to elaborate on this concept by referring to specific archaeological examples, including cognitive science, philosophy, theoretical archaeology etc. Now it is time for the concluding discussion of the most important issues, or more specifically those which seem most important to the author of this work.

When thinking about the reasons for the dynamically growing social differences in the Neolithic period, which stood in contrast to the stagnation of the Palaeolithic era, Ian Hodder also paid attention to the role of material culture. Rather than slowing the pace of social changes, material culture accelerates them which are, according to him, a paradox that require some explanation. It might seem that the more objects there are around us the more difficult it is to change the human relations that are "filled" by them. In reality it is the complete opposite!

According to Hodder, the growth of material culture was accompanied by an increase of its "objectification". While creating a social structure towards the outside, it underwent an objectification in the world of human craftsmanship in separating from the subject, and due to this, it could be worked upon. Along Hodder's train of thought we might add, by paraphrasing Bruno Latour, that a greater number of objects require a greater amount of subjects, and a greater degree of subjectivity requires a greater level of objectivity[1]. What can this mean?

It seems that the meaning of the interpretation mentioned above can be figuratively expressed by introducing the reference frame concept being material culture itself. Instead of locating social structure into a timeless mythological world it becomes somewhat materialized and perceived as a

---

[1] Latour 1993, 108.

sphere, which can be constructed continuously anew, presented on the outside in the form of objects which create–as Jacues Lacan would call it– a "phantasy frame"[2]. As we know, such a frame is a sort of paradox as it is a limited field of unlimited possibilities of expression. The frame takes on the role of language which, thanks to its recurrence (a basis for later computability) seem to offer unlimited means of expression as well. This materialization assures man in the belief that he can change reality, and so by manipulating objects in social relations, humans began to manipulate society in a way never seen before[3]. The greater the number of objects, and the concepts and metaphors based around them, the more manipulation occurs, whereas a significant role is played by the development of concepts and metaphors connected with shaping the mathematical mind, which I have tried to present in this work.

The key issue of larger human groups is trust. As human society is built on the concept of cooperation and sharing resources, trust becomes the basic factor for survival. Therefore when at the end of the Palaeolithic period human groups began to grow, in time exceeding the capabilities of the human brain to comprehend social relations as a whole[4], it became necessary to develop other communication media, which extended, or even substituted, genetically limited abilities. Founding a new idea of society, whose main visualization was settlements and their architecture, became necessary[5]. It was architecture that was given the special role in manipulating society, which is convincingly pointed out by many authors[6]. It has the power to speak to the whole of society both in a single moment and all the time. Hence, the organization of Neolithic settlements and the architecture of buildings could constitute something similar to "theatres of memory" as well as become a reference for further actions[7].

When referring to Hodder's thesis we may also expect that the development of metrological–mathematical thinking had an influence on the increase of objectification and the rate of the socio–economic dynamic. We may now define this process as a metrologization of material culture

---

[2] After Žižek 1992, 171.
[3] Hodder 2004, 45–52.
[4] We, modern humans, are constrained by our evolved cognitive ability to handle the complex web of social interactions of up to about 150 people, but in common with our southwest Asian early Neolithic predecessors, we habitually live in social groups that far exceed that cognitive limit.
[5] Watkins 2006, 15–24.
[6] Lewis–Williams & Pearce 2009; Watkins 2006; Bourdieu 1973, 98–110; Lévi–Strauss 1991, 434–436.
[7] Watkins 2004, 5–23.

by the use of two basic metaphors: the collection metaphor, which had a distinct influence in the Near East, and the measuring stick metaphor, which found better grounds in Europe. It allowed more efficient communication in the sphere of material production, maintaining the trust of partners involved in the exchange network because metrological and mathematical concepts, when released from the timeless narrative sphere, enabled the establishment of an additional platform in social relations, anchoring on the understanding of rational rules of proportion and simple arithmetic. If the assumption that macrolithic tools were also perceived in a numerical (proto–numerical) way, as I tried to present earlier, is correct, then we would be faced with a development of rational communication methods. Macroliths were a medium of communication which fulfilled many conditions of rational communication that were proposed by Habermas [8]. For example, although macrolithic blades were highly specialized tools in the Eneolithic period, their exchange and fragmentation could also create new forms of expression, new messages, which used the "excess" of a numerical or quasi–numerical content (compare chapter 4). They could function as recording systems or as symbolizations of the rank of their owner by the use of grading the blade size. There were probably many more possibilities and we will not be able to discover them all. In my earlier studies I tried to show that the development of the metrologization of a material culture could have even influenced the creation of a new vocabulary and metaphors in social relations[9]. As the observations of the macrolithic industries suggest, this new vocabulary was used to emphasize the social inequality. A figure, which I humorously dubbed Mr. Blade[10], became the central person of some European societies of that time. In that person's mind the measuring stick metaphor found good material for further development.

It seems that, due to the above, this four–element metrological structure, which emerges from the analysis of the macrolithic and copper bead industries, was not created by coincidence[11]. We would be faced with the starting point of early counting and measurement systems which had their origins in the development of material culture. Here I would like to call upon Greenberg's observations. He stated that the counting base in simple social structures is often analogical to the metrological systems existing in those structures. The collection metaphor can be fitted to any chosen group or products as part of the development of mathematical

---

[8] Habermas 2002, 481; Dzbynski 2008, 103.
[9] Dzbyński 2011; 2011a.
[10] Ibidem.
[11] Por. Dzbyński 2008, 205.

perception. In other words, the collection metaphor may lead to a cognitive identification of full collections–filled containers, whole tools– with units of measure. Greenberg's conclusion is supported by data from South America, where the words used to describe large numbers, such as 4,000 or 8,000, are derived from words which define different types of vessels[12].

If we assume that numbers were initially treated as vessels containing objects then working out the concept of the counting base would correspond with the filled vessels[13]. And if numbers are treated in the same way as vessels or special tools then the counting base may derive from full vessels or complete tools. In regards to the above, the thesis, according to which the measuring stick metaphor is also responsible for the creation of the first counting systems in Europe, imposes itself automatically. This thesis is all the more probable as there is no evidence of any use of the *digit* metaphor in Europe which was an element of a collection of objects or a base for the development of notation systems in the Near East. Tokens were in fact a materialization of that metaphor.

A good example for the methods of thinking and acting described here are metrological structures in the Celtic languages described by Justus[14]. It was a counting system of a proto–numerical character with a mobile base of 4 or 5, which according to Justus is a consequence of the metaphorization of filled vessels. The mobile base is also interesting because it illustrates an exceptionally easy transition from a quaternary system, which is visible in the analyses of Eneolithic materials[15], to a quinary system, which is characteristic for the Bronze Age[16]. The expansion of the counting base did not have to be a revolutionary change but a smooth one. Moreover, Justus confirms that the ancient numerical and metrological systems were largely identical and were based on ranking units, which had a specific place in the sequence (relational). The words which defined those numbers/measures, such as *lethera* and *methera* retained their prenumerical linguistic forms similar in their meaning to the English "half again as much" and "next to the last". These words did not have any abstract numerical meaning, but nevertheless could be used to perform simple arithmetic operations.

Although the human body is the basic source for the development of mathematical perception (e.g. counting with fingers), along with the

---

[12] After Justenson 2010, 45.
[13] Ibidem.
[14] Justus 1999, 55–79.
[15] Dzbyński 2008, 205.
[16] Sommerfeld 1994, 238.

increase of material production, the centre of mass is moved to material objects. The establishment of the first metrological systems based on the mathematical proportions rule as well as the forming of a counting base seem particularly important in this case. These phenomena had developed in a body–object–society relationship while the significant motivation in expanding metrological/numerical structures was the expansion of the sphere of production and exchange[17]. Moving the idea/metaphor of the measuring stick from bodies to sets of objects meant an escape of the measuring concept from narrative contexts with a strictly specified subjective identity, in which those bodies functioned, to the context of objectified social relations, where from that point they could develop more dynamically and from that came into other niches of human communication. Objects laden with the number concept began to emerge from the timeless mythological/narrative sphere. The fact that the process of technical development is present on more than one occasion here is not coincidental. What does this mean?

Allowing myself to digress, I would like to invoke Heidegger, who believed that gods cannot be present in technology[18]. Technology is the end of metaphysics and the beginning of an advanced way of thinking and acting. Naturally, technological development may be filled with religious content, as many authors point out, but only in its secondary discursive form, such as in the case of the idea of technological development as a path to the salvation of human kind of contemporary pre–modern Europe[19]. Sooner or later a cognitive turn similar to the one perceived by Bruno Latour in the vacuum pump experiment which took place in the 16th century, and the discourse surrounding this phenomenon, occurs. This turn is initiated by the mechanism of introducing measurements, numbers and technology into the sphere of social communication, which leads to a significant cognitive dichotomy. People begin to perceive themselves as completely different, using different languages, as for one group a higher being ceases to be necessary to understand the world[20]. In my opinion, the prototype of this cognitive turn was the craftsman/craftwork specializations of the Eneolithic period, in the framework of which the liberation of a new way of perceiving the value of metal and flint in metrological categories had occurred. Hence, we may say that it is only with the Eneolithic societies that the actual parting with the Palaeolithic language and tradition took place.

---

[17] Kula 1986; 2004.
[18] Heidegger 1982.
[19] Noble 1999; Sawday 2007.
[20] Latour 1993, 111.

To present the above statements in an appropriate light we must return to Lakoff's initial thesis. Both the near–eastern tokens as well as the European measuring stick were also metaphors–links between the body and material reality. Tokens were probably symbolic manifestations of fingers–*digits*. The measuring stick also had to have its origin in the human body, which is clearly suggested by the hypothesis of a measure of length existing in the Neolithic period: Thom's yard, Rasch's Neolithic length or Pavuk & Karlovski's fathom[21]. It is the modern equivalent of the royal cubit or foot, which we know from ancient times. However, in both cases we are faced with an identical model situation, which enables the making of the concept of numbers and measurements more abstract, but in a slightly different way. Tokens were related to physical goods, while the measuring stick referred to rich narrative content connected with architecture. It was only with its migration to the sphere of intersubjective actions connected with production of tools that allowed for a fuller release of its potential to rationalize communication.

I wrote elsewhere that initially the macrolithic industry as well as early metallurgy technology took up identical areas of communication and exchange in Eneolithic societies [22], which is after all accepted by archaeologists who point out the fact that both technological paths are similar[23]. Macroliths were a sort of *alter ego* of metal in a time when metal was preparing society for its complete acceptance, with which many economic, ideological, cognitive and other consequences were connected[24]. In my opinion this implies the following assumptions: both macrolithic technology as well as metallurgy initially made use of the vocabulary and cognitive patterns available at that time whose base was the measuring stick metaphor and spatial measuring which had its earliest use in architecture. This presents us with the leading Eneolithic technology as the *spirytus movens* of advanced means of mathematical cognition. Where metallurgy technology had been developed, societies achieved a certain cognitive turn based on the conceptualization of the abstract weight measure, while there was no metal–they remained on a more traditional development level.

That is why I do not think that blade fragmentation was an insignificant phenomenon when taking prehistoric communication processes into consideration. The theory which supports the enchainment

---

[21] Karlovsky & Pavuk 2002, 137–156; Rottländer 1999, 189–202.
[22] Dzbyński 2011, 172–184.
[23] Strahm 1994, 315–327.
[24] Eliade 1993; Kowalski 2000, 205–210.

of relations in the Eneolithic period does not explain it completely[25]. In fact we are also faced here with a process of forming early mathematical concepts, which were a sublimation of language communication acts to a numerical form which took place over tens of generations. They were most probably present in language as expressions of a specific grammatical structure, all but impossible to guess today, which were the basis for further mathematical reasoning and development in Europe. **In other words macrolithic tool manipulation in its most effective form could serve as a reference to forming the first counting systems in Europe. It was a system which made use of recursive grammar, in which the metrological concepts were probably identical to the words which described numerical values.**

In the Near East strange counting bases, such as 12 and 60, had developed. Their creation was probably connected with the control of greater and more various resources in the staple financial system [26] (compare below) within a framework of a much more versatile society. This was not the case in Europe. The four–element structure coming from the manipulation of the measuring stick, the basic metrological concept of Europe was sufficient to operate metrological communication in Eneolithic societies, in which individualized partnership obligations were the main background for exchange. In Europe no evidence has been found so far of the existence of any control of goods in the framework of obligations towards a community[27].

John Barrow, a mathematician, supports this hypothesis. In his studies concerning the development of mathematics he pointed to the fact that in all Indo–European languages numbers above 4 are never treated as adjectives which adapt their form to what they describe. Whereas numbers up to four and including four, are subject to declination. That is why we can say: double, triple, quadruple but not "fiftle". This is because it sounds odd. According to Barrow, this linguistic phenomenon shows that the relations between the noun and adjective are well–established and that it is a very old structure. Moreover, Barrow also referred to ancient metrological systems pointing out that the number four played an important role in them. Such a system could have extended quickly to the 8 counting base and then 10, one piece of evidence of which is the number 9, which in some languages is connected with the word "new": *novus–*

---

[25] Chapman 2000.
[26] Brumfiel & Earle 1987, 6.
[27] We have to recall that tokens appear in European social contexts where archeological evidence shows the existence greater and more varied societies.

novem, neu–neun, etc[28]. However, Justus points out to the fact that the transformations concerning methods of counting and measuring, which originated from a 4 base, operated throughout history without much difficulty, generating even a vigesimal counting system, which spread through Europe and is present to this day in the Basque language[29].

Although they were probably more complicated, through comparing the paths of this development with archaeological evidence we are able to say with a great probability that the quinary or decimal system was present in the Bronze Age in a part of Europe, to which the inter alia analysis of symbols found on sickles, presented by Sommerfield, point to[30]. The earliest certified written counting system in Europe was the decimal system present in linear writing[31]. However, the existence of a decimal system in the later Bronze Age does not exclude an earlier version which consisted of four or eight elements as its initial counting base in Neolithic/Eneolithic Europe. Such a possibility was suggested recently by Shan M. Winn when taking into consideration studies concerning the mysterious Balkan scripture[32]. In my opinion, regardless of the detailed results of the research we are now able to anchor the development of early mathematics in the material culture of European societies, which I have tried to show in this book.

Between Europe and the Near East

We began the study of rationalized methods of communication with the example of Near Eastern tokens (chapter 2). I think that at this point we should ask ourselves why this development was so different in the Near East and Europe. In Europe it was not the tokens but manipulations of the measuring stick metaphor, initially in the architecture and then in the macrolithic industries that led to the comprehension of mathematical rules. European tokens played some role in communication processes but their range was too limited and we know too little of them. Below I would like to attempt to make an interpretation of these two divergent development trends–the Near Eastern and the European.

The characteristics of the architecture of Near Eastern cultures are already visible in such sites as Nevali Cori or Göbekli Tepe. In the latter, the early megalithic structures, whose size is impressive, with almost 10

---

[28] Barrow 1999, 71.
[29] Justus 1999, 70.
[30] Sommerfeld 1994, 236–245.
[31] Anderson 1958, 363–368.
[32] Winn 2009, 49– 62.

feet high T–shaped stone pillars were based on a circle plan and decorated with animal representations. A researcher of that site–Klaus Schmidt–has no doubt that the pillars represent shamans, because the early structures were raised by hunting communities[33]. A much later scene from Çatal Hűyűk, on which we may see an image of human beings decapitated by winged creatures shown as through an X–ray, seems to confirm this interpretation[34]. First, the ritual spot was round but later it became square as in Nevali Cori. It was separated from the outside creating a sort of shrine which was unavailable for any observers. The up–down directions, verticalness in perspective linked with the *axis mundi* concept, had been underlined in these structures.

Round buildings remained a common testimony of the central concepts in the Near East, even in the Mesopotamian civilization period. They often have an administrative function, which has been noted in Tell al–Rapa[35]. Round buildings were placed in the settlement centre, the rectangular houses being adjacent to them. However, as Lewis–Williams and Pearce point out, elements which emphasize verticalness are emphasized everywhere. They mark their presence in Göbeki Tepe, they are in Jericho and they are subject to specific adaptation in the Çatal Hűyűk architecture, where the only way to enter the buildings was by use of a ladder[36]. The dead were most often buried under the floors, pillars and columns were raised as well as platforms, which were placed at different heights, appropriately painted, decorated and marked–all these behaviours emphasized the verticalness rule. It is also seen in the so called temples in Çatal Hűyűk, where pillars had been crowned with images of bulls (*bucrania*) separated by scenes of mythical meaning, which Mellart interprets as a mythical hunt[37]. The verticalness rule is present thousands of years later in the form of Near Eastern ziggurats, symbols of a sacred mountain, where only the highest priest was permitted to approach the highest spheres of sacrum. Meanwhile, the horizontal division was connected more with social control, forming structures according to which the division into participants and outer spectators was established[38].

The later concepts in Göbekli Tepe, introduced in the period of the advanced Neolithic period, illustrate the dramatic change. T–shaped pillars shrunk to 1.5 meters in height and were placed in an appropriately small

---

[33] Schmidt 2006.
[34] Mellaart 1967.
[35] Schwartz 1994.
[36] Lewis–Williams & Pearce 2009, 110.
[37] Mellaart 1967; after Lewis–Williams & D. Pearce 2009, 116–117.
[38] Lewis–Williams & Pearce 2009, 102–110.

rectangular–shaped building. The initial sacred circle had been replaced by a rectangular temple. In Europe, however, this history seems to proceed in the other direction. The square idea, connected with the crossing of lines and the connection of straight lines, seems dominant at the beginning. Only in time does the circle idea gain an advantage over the square in Europe, which is illustrated by such ideas as the Eneolithic enclosures as well as Stonehenge.

One of the best studied Neat Eastern settlements, which offered an insight into almost 5,000 years of construction tradition, is in Çayonu which is located in the south–eastern part of Turkey. Çayonu has supplied strata from all Neolithic periods, from PPNA to the ceramic Neolithic and as a result we may observe the most important trends in construction. In the first stage, round or oval constructions were built with their floors rammed into the ground. In the subsequent stages rectangular constructions on a so called grill plan, which probably were the foundations of a higher floor, started to be built. It was only in this stage that constructions had a similar shape and orientation and placed according to a chessboard design. These regularities point to the fact that a general plan for the settlement might have appeared among the PPNB which was based on subjecting all constructions in one direction, to be followed by all. However, in the course of the next stages, constructions divided into small cells connected with Neolithic farming and storing goods, were being built. Constructions were not being oriented according to one direction but it was noticed that the largest and best equipped ones were concentrated around a central oval square, which was named *plaza*. In subsequent stages round constructions of cult and administrative purpose had been built. Eventually, in the ceramic Neolithic phase the settlement transformed into an agglomeration of constructions close to each other, separated by narrow irregular alleys, which became the characteristic feature of cities in the Near East [39]. The settlement agglomeration concentrated around a temple which towered over its surroundings.

The settlements became more and more complex, but it is important to notice the specifics of this process. Although they had to be planned, Near–Eastern settlements do not show such a connection to the linear setting as their European counterparts. Each household was a sort of separate social cell, similar to a bee hive, which was underlined by small walls and fences or placing house entrances on the roof. Rectangular

---

[39] Gates 2003, 20–21.

houses were very often built onto one by one creating tight irregular settlements just like that in Çatal Hűyűk.

Thus, as Clive Gamble states, the same theoretical perspective can be seen in the evolution of Near–Eastern architecture as in the material culture in general. Referring to Chapman's enchainment–accumulation concept, Gamble suggests that the same should be expected from the architecture of settlements. Due to their systematical growth structural restrictions connected with access to buildings and their visibility, which at the same time limited the autonomy of each household appeared. They became a barely visible part of the whole settlement. A fragmentation of space, which is an enchainment of separate household units as well as parts of the households, which in turn led to the creation of compact, or agglomeration settlements, can be seen there. Moreover, each house was also a container–in the metaphorical sense–in which materials were gathered (accumulated), sometimes in amazing amounts. That is why, according to Gamble, this process should be linked with another great Neolithic invention–the mass use of pottery. The trend towards establishing small settlements, which proceeded along with the spread of ceramics, could have been understood as a symbolic implementation of the container/capacity metaphor in ceramic vessels, which became the customary equipment of the deceased buried in cemeteries who were separated from the settlements. This form of container–the ceramic vessel had an almost infinite potential of divisibility and reproduction as vessels could be made without greater restrictions[40].

According to Wengrow, the increasingly meticulous division of houses also suggests a growing dichotomy of gender and its assignment to specific economic spheres of life[41]. Round constructions were most often connected with female activities while square ones–with the male control of administrative activity. Hodder saw a fundamental ideological transformation in this process, during which women were initially identified with wildness and nature, which is visualized by images of a goddess sitting among ravenous cats[42]. In the end, however, men performed a symbolical takeover of the wildness, taming it by the use of the idioms of the warrior, hunt and communal drinking, decreasing women's role to that of an addition to the fireside. The final stage of this development was, according to Wengrow, the introduction of the trisection of the Near Eastern household, also expressed symbolically–as one of the metaphors which brought order to social life. The trisected building

---

[40] Gamble 2004, 91–92.
[41] Wengrow 1998, 783–95.
[42] Hodder 1990, 5.

became a base structure of household activity and as a farm metaphor it was extended to the administrative sphere, the sphere of production and ritual activity, which influenced the further facilitating of new ethics and increase of productivity[43]. This occurred in the Ubaid period, between 5,000 and 4,300 BC, directly preceding the dynamic development of urbanization as well as the booming development of complex tokens.

During the time when Near Eastern settlements became more structuralized and internally limited in the form of agglomerations, tokens took the role of the main medium of communication and social control. It was also a period in which, as Wengrow states, regional cultural features ceased to matter. New media of communication (tokens, stamps, proto–writing tablets) created a new social awareness[44]. Soon palaces and temples became the emanation of wealth and truth, they took over the control tasks, scripture began to be developed. Tokens were most importantly a method of controlling the economy and social activities, which was the domain of specialists for a long time. The mathematical concepts which they represented were connected with obligations towards the community represented by social leaders. Instead of a complex manipulation of architecture, which appeared only during the period of the rise of civilization, recording systems took a significant part of social control mechanisms on themselves.

In Europe the initially loose building type in an early Near Eastern Neolithic style changed into settlements built on a perpendicular line plane. The role of cosmological connections, which those lines represented, was to socialize society; they were also an equivalent of Near Eastern shrines. They were an expression of a community of beliefs connected with the sphere of social order. Further in the north such a function was taken over by long houses. The line metaphor earned a prominent place in the social order symbolization in Neolithic Europe, which had been subsequently materialized in the form of megalithic tombs as well as enclosures, which despite their round shape were inscribed into crossed straight lines, having undoubtedly a cosmological meaning. On the other hand, the tokens present in Europe had been integrated with ritual communication systems, although the role they played amongst them is not perfectly clear[45].

Hence, it seems that in Europe we are faced with a different distribution of accents, that is with an emphasis of the linearity rule, which was visible both in south–east European settlements as well as in houses in

---

[43] Wengrow 1998, 790.
[44] Wengrow 2010, 55.
[45] Budja 1999, 219–235.

the north, and finally in the remaining monumental cult concepts. The concept of *axis mundi* probably did not disappear (it exists in the form of mandala type symbols or the spiral), but the architecture *en mass* was subjected to it in a different way than in the Near East, which emphasized the horizontal system of lines. This is perfectly visible in Newgrange where falling sun rays on a midwinter's day on a spiral–ornamented slate located at the end of a corridor express cosmologic connections [46]. However, not only long monuments and tombs are subject to the power of the line. As Thomas points out, also round monuments, such as Bryn Celli Ddu or Stonehenge, have definite linear adaptations[47]. It seems then that in Europe stress had been put on lines, reflected on the surface and having a cosmological meaning. They were a means for uniting the cosmos with architecture and in this way also with all of society, not by emphasizing the vertical but the horizontal, which in my opinion is a sign of the European way. Lewis–Williams and Pearce identify a whole group of elements whose purpose was to achieve this unification: positioning the corridors towards the sun, which indicated key natural events; the entrance to a tomb; the sun's rays which fell along the corridor; the placing of a megalithic spiral symbol[48].

The emphasis put by Lewis–Williams and Pearce on the similarities between Near Eastern and European architecture at the same time show the subtle differences between them. Everywhere architecture was an embodiment of society and/or the cosmos, but not everywhere was this achieved by the same means. In the Near East, apart from some exceptions, no directing of architecture towards some specific points on the horizon was observed, which is the basis of its features in Europe. This is undoubtedly an important difference. In Europe architecture was often an open manifestation of linearity (long tombs), although in time it became more complex, and its connection with distant points on the horizon.

We must also remember that a metaphoric understanding of linearity in Neolithic settlements in Europe created a symbolic level which connected two worlds: the cosmic and the earthly. Both infinity, which connected settlements with the cosmos as well as finitude, in the form of the first measures used when building enormous constructions were manifested in this linearity. We may expect the same in the case of megalithic constructions. Hence, the linear measure could have been an ideal

---

[46] Lewis–Williams & Pearce 2009, 231.
[47] Thomas 1991, 45.
[48] Lewis–Williams & Pearce 2009, 242.

rationalization of social discourse as it was idiosyncrasy, filled with great mythological and social meaning alike.

Although the transformations mentioned were probably not a simple and linear process, we may outline two poles here. The communal aspect in the form of architecture was connected with the mediating role of straight lines in the monumental form and it spoke out to the whole of society at the same time. Macrolithic tools became the other pole, which were prestigious objects and which emphasized the individual aspect. The connections between these poles are multidimensional. What I wish to say by that is that Eneolithic tool making was not only a case of borrowing or earlier tool traditions but they show some connections with architecture. Macroliths are a product of the period in which the idea of linearity or, as some would rather call it–the metaphorization of the line, celebrated its greatest successes. That is why I believe that we cannot separate these two phenomena.

Explanation attempts

While thinking of the causes of such a different course of evolution the concepts of measure/number in the Near East and in Europe, let us consider the natural conditions. In the Near East the Neolithic period developed gradually on the base of hunter–gatherer cultures which lived there for a long time. In fact, as some authors point out, the basis for the development of a farming mentality appeared very early, achieving the so called cognitive fluidity stadium early in the Upper Palaeolithic period[49]. "Abandoning of the caves" at the end of the Ice Age resulted in the development of symbolic architectural structures and the construction of those "memory theatres", in which the cognitive centre became perhaps the *domus* idea as a basic medium of control: the house as a ritual centre, a grave for ancestors and shelter. Meanwhile, the environmental conditions necessitated the managing of a surplus. Irrigation required group cooperation which resulted in future crops. This could have favoured the development of some reciprocal expectations, whose objects were in fact farming goods[50]. This would explain the development of tokens which best fulfilled the control tasks. Whereas in Europe the environment was all the more favourable for farming that it did not require the energy investment of the whole of society. Hence we are presented with two different notions, the Near–Eastern one which puts pressure on group social control while

---

[49] Mithen 1996, 171–210.
[50] Bellwood 2005, 65.

the European notion on the development of individualized exchange relationships. Although the constructions of houses, settlements and megalithic graves required the cooperation and effort of the whole community and they were an image of that effort, the prestige object exchange relations, which were the other pole, were not a sphere of dominance of religious specialists but rather fitted technical specialists and community leaders.

The Near Eastern recording systems concerned the most basic farming goods (grains and animals), which were the staple financial sphere[51]. Moreover, Schmandt–Besserat stated that among tokens, only on rare occasions could those referring to luxury goods be found, although this statement is not completely true. On the other hand, European exchange systems, in which framework fragmentation had been practiced, concerned mostly prestigious, luxurious (macroliths, copper, even spondylus) goods or goods of a special purpose (figures, vessels). The mechanism of control was an element of the Near Eastern exchange solution from the very beginning, while in Europe individualized relations towards partners were the basic level. Moreover, it is beyond doubt that the path of European development was less varied. There was no need for the control of all goods, only those few ones crucial from the perspective of traditions or the current trends. Nevertheless, both in the Near East (tokens) as well as in Europe (macroliths) we are definitely faced with the process of the development of communication media which rose above locality[52]. This fact allows us to examine these phenomena, distant from each other, on a mutual theoretic level.

That is why the model of social evolution presented by Kristiansen[53] is a good supplement to the above. Referring to the world system concept he proposes to see the process of social development in a two–trajectory perspective. The European way, connected with the control of prestigious object exchange (wealth finance) was characterized by its decentralization and the dissolution of authority, the consequences being the appearance of proto–feudal structures in the Bronze Age. On the other hand, the Near–Eastern way, which put pressure on the control and redistribution of common products (simple finance), resulted in the creation of theocratic states.

At the end of this sub–chapter let us think about connecting the above mentioned theses into one concept, which could be an answer for the question in what way architecture could constitute a cognitive base for the

---

[51] Brumfiel & Earle 1987.
[52] Wengrow 2010, 55; Budziszewski 2006.
[53] Kristiansen 1991, 16–43.

varied adaptation of macrolithic flint industries in European social structures?

As Watkins suggests, Neolithic architecture was a completely new social creation. At the same time, being somewhat a basic generative form, architecture could have also supplied a quasi–grammatical structure (also on the vocabulary level) in which a framework of symbolic forms and objects as well as ritual activities could be realized, being a part of a "multimedia drama"[54]. On the other hand, as studies performed by Lewis–Williams & Pearce show, lines, spirals, zigzags, points, etc. were the main elements of this new grammatical structure as well as metaphoric references[55]. All these elements contributed to architecture, although they were manifested differently.

Perhaps as an element of the Atlantic tradition, that is circularity, clearly had its mark on megalithic structures. The time has also come to perceive linearity and its extensive metaphoric and semantic relations with architecture, the landscape and subsequently with an individual person of the Eneolithic period. In order to fully understand this matter I shall refer to a historical tradition.

The process of leaving the age of myths by the Greeks resulted in a vision of the Earth being more or less round, which probably originated in the Near East, as Martin West suggests[56]. Later philosophers supplemented this vision with a metaphor of the world as a circular sphere which became relevant for the next 2000 years, although it was a subject of numerous modifications, especially in the Middle Ages[57]. According to Plato, the spherical shape was the perfect shape, while the circular motion was the perfect motion[58]. Ptolemy supplemented this image with the system of "Ferris wheels", as Koestler dubbed the first theory of epicycles[59]. However, in the end classical ontology remained sphereology, it concentrated on "what is regular, circular, spinning–in–itself", it was a logic, aesthetics and an ethic of round things, as Peter Sloterdijk recently summarized this mental development[60]. According to him, the circular shape could have worked as a "cosmic immune system", a metaphor of man's unity with the cosmos as well as a universal communication language with the creator. Moreover, Sloterdijk seems to point out that the

---

[54] Watkins 2004.
[55] Lewis–Williams & Pearce 2009, 210–223.
[56] West 1999, 143–150.
[57] Tarnas 1993.
[58] Plato, The Republic.
[59] Koestler 1968, 66.
[60] Sloterdijk 2005, 18.

sphere metaphor was quite quickly and efficiently the subject of a transformation into a crystal palace, a technological greenhouse, an idea and practice of globalization which offered a sense of wealth and safety for humans of western culture.

A truly interesting fact when considered in the context mentioned above was that the European cosmologic visions were from the very beginning far more geometrized than the Near Eastern ones. The beginnings of mathematics in Greece are inseparably connected with Pythagorean geometrization, in which framework the "number–shapes" were initially contemplated during the process of discovering laws of the universe[61]. The question concerning the beginning of this tradition in connection with pointing out the differences of both regions mentioned earlier seems an intriguing subject.

However, let us return to prehistory. The stick or line metaphor can be easily manipulated as it is a dynamic and *de facto* highly abstract. It may manifest itself as an *axis mundi*, a long monumental tomb, a lineal measure, a human body or the idea of infinity. Such a broad spectrum of the possible uses and meanings of this metaphor makes its change, as well as its sublimation accompanied by a rise of abstraction much easier. Thus the measuring stick idea could separate itself from the narrative sphere without effort and fulfil more and more specialized communication tasks. The element, which earlier was inseparably linked with the wheel, spiral or squiggle as an embodiment of cosmic connections was quickly put to use as part of individualized communication acts. To put the issues discussed here in the form of a metaphor and analogy, **I would say that if the *domus* idea was a revolutionary doctrine in the Neolithic period, then the measuring stick was its executive act.**

As the measuring or simply stick/line metaphor had greater rationalization potential than e.g. zigzags, spirals, points, etc., the grammar and vocabulary connected with constructing architecture could encompass greater areas of activity, among which macrolithic tools seem to be its adequate extension. The measuring stick concept, apart from perfectly fulfilling its role in forming new anthropogenic places, connected with a sedentary lifestyle, could have been a source of production for further metaphors connected with tool activity. Archaeological sources, as well as the interpretations made so far, clearly point to the fact that architecture and macroliths were not conceptually separate, but this should also be understood as a transmission. Apart from its influence in constructing the anthropological landscape, the linearity metaphor and the measuring stick

---

[61] Koestler 1968, 31.

metaphor also encompassed phenomena which from today's point of view seem distant, that is idiosyncratic tools, such as macroliths.

As we can easily see, this concept concerned language, vocabulary and metaphors. By relating this matter to the present, Richard Rorty assumes that in the case of creating new vocabularies it is often hard to identify a single source[62]. However, the world was far simpler during the transition from hunting to farming. The world of Palaeolithic hunters was dominated by images of animals with a small number of human figures, which is undoubtedly reflected in Palaeolithic vocabulary and metaphors. A dramatic change occurred along with the Neolithic period which was based on introducing new metaphors into this vocabulary. These were connected with the dynamic development of material culture, among which architecture played a leading role and in the next stage–new technologies[63].

Value and accumulation

The key for understanding the phenomenon of social inequalities is the problem of the value of material goods and the possibility of their accumulation. These are also phenomena which are strictly connected with the ability to count and measure. All the former agreements made by prehistorians clearly show that the hunter–gatherer societies which lived in the Palaeolithic period were basically free from the accumulation of goods. The value of material objects was undoubtedly crucial, although it seems that at that time it had a distinct functional and symbolical–mythological[64] (narrative) accent, which did not favour the development of material differences. It was an egalitarian society, which means one in which individual differences changed quite smoothly as it had support in the personal achievements of each person. A successful hunter was always valued, as well as a successful shaman, although this did not give an established position in a certain group for a lifetime. Throughout a great part of human history a position inside a group was set dynamically according to ones contribution to the community[65], wherein the basic reference for the hierarchy could have been family bonds[66].

The situation began to change with the coming of farming. Social attention rested on the non–functional value of objects while human

---

[62] Rorty 1989.
[63] Dzbyński 2011a.
[64] Mithen 1996, 192–195.
[65] Barkow 1975, 553–572.
[66] Lee 1978, 871–895.

relations took on a greater distance, which had been successively filled with different objects and goods signalizing social differentiation[67]. The quantitative and qualitative (metallurgy) technological development led to an enhancement of the relationship between the high value of objects (measure) and the high value (rank) among people. A stress had been put on artefacts, which could express (symbolize) the meaning of human relations. In other words, artefacts became identified with friendship, authority, obligation, social rank, etc[68]. Some of them, in the light of Habermas' ideas, carried medial aspects in that they were countable and measurable[69].

The climax of this development was money. According to Renfrew, the introduction of money in the form of coins was such a momentous event in the history of mankind that we may treat it as a caesura between prehistory and history[70]. Of course money is only an accumulation of the process which gained momentum throughout several thousands of years, since when objects (axes, macrolithic tools, etc.) had been exchanged as valuable and rationally measurable objects. This period is still only dimly explored, nevertheless–this is certain, it was essential for the forming of such a value concept and such a social anchoring, which is very close to us today. For this is a period in which numbers/measures became connected to objects and in this way were introduced into human relations[71].

There is no doubt that the clear caesura concerning the formation of the modern concept of value is the invention of metallurgy. Despite the fact that archaeologists use the term "rich" when describing some graves, in reference to earlier periods, it was only the invention of metal which introduced a completely new quality. Naturally, the differences in grave possessions occur as early as the Neolithic period, but it is generally difficult to speak of ostentatious burials, which are a characteristic element of the Metal Age. Ostentatious, grand burials appear along with the development of metallurgy and one of their earliest forms in Europe is the cemetery in Varna, near the Black Sea. Varna is the first site where the discovered graves contained rich decorations, tools and other metal objects. Moreover, here we are faced for the first time with significant amounts of gold, a material used for making objects of no utility value. The presence of copper in the Varna graves should be examined in the same way, as only after the invention of hardening methods could this

---

[67] Hodder 2004.
[68] Schiffer 1999.
[69] Habermas 2002, 481; Dzbyński 2008, 103.
[70] Renfrew 2007, 161.
[71] Dzbyński 2008.

metal be used for making practical tools. The massive copper axes present in those graves could only be a means to show the accumulation of the desired substance. In that case it can be said that in Varna we see a transformation of valuable and useful objects into purely objects of value[72]. According to Chapman, in Varna for the first time we are faced with the phenomenon of wealth accumulation. The specific role of metal as the material which embodies this accumulation is also underlined[73].

The richness of metal in Varna acquires a particular tone if we perceive it as the first wave of accumulation of such strength, which decreased in time or was redistributed as Staaf's analyses suggests (chapter 5)[74]. Metallurgy technology was still in its initial development stage. Copper could not be processed to harden it, casting was performed in simple open forms after which the metal was processed by means of forging, sanding and polishing techniques known at that time. Ornaments were made from thin copper or golden plates. This first stage of technological development makes us wonder if metal had the same meaning for these communities as for those in the Bronze Age. Was it the subject of the same conceptualization? Weight is an adequate means of assessing the value of metal. Was it used in the times of the Varna culture? We may have some doubts as to that. Consequently, we may also have doubts as to whether metal could be measured rationally.

If at this early stage metal was not conceptualized in categories of abstract weight then how was wealth and prestige measured? At this point we need to recall the macrolithic blades which appear somewhat on the margin of the Varna burial grounds which were rich in metal. Their presence seems to be a vital supplement of an ostentatious burial, taking into consideration that they were a concrete number representation in the form of the measuring stick in a time when the abstract concept of weight was unknown. The fragments of blades were put into poor graves while the longest blades were placed in the richer ones. Thus blades were a symbolic expression of wealth which used the stick/line metaphor, and in this way referring to the old traditions of the social order expressed in architecture (compare chapter 3). This is symptomatic in one of the richest graves in Varna, which contained 255 golden objects, among which the longest blade, 43.3cm long, was found[75]. We may say that these blades were as long as all the metal accessories deposited in the graves. Although they were a product of advanced craftsmanship (similarly to metallurgy),

---

[72] Renfrew 2007, 168.
[73] Chapman 2000, 47.
[74] Staaf 1996.
[75] Pellegrin 2006, 37–68.

they manifested the presence of tradition clashing with a new phenomenon of accumulation. That is why Varna remains a fundamental image of the confrontation between the old world and narrative values still of Palaeolithic provenance and the world of accumulative value as the foundation of civilization.

The process of fragmenting blades as well as the manipulation of macrolithic tool length in different social contexts seems to be a certain anticipation of the actual wealth accumulation phenomenon, which characterized the Bronze Age at times when developed and abstract methods of mathematical perception had been used. Before this occurred, we must perceive the earlier stage as the forming of such rational methods. It seems to me that the missing link in the debate so far is the fragmentation of macrolithic tools, a fuller understanding of which I proposed a new dialectic method in chapter 4. If we agree that Chapman's theory is the basis, according to which the "enchainment" of social relations through the fragmentation of objects was the main method of communication in the Neolithic period, then the fragmentation of macroliths appears not only to be an antithesis, which introduces mathematical/metrological concepts into the relationships of this "enchainment", but also a transitional stage between the lack of a numerical description of valuable objects and their highly abstract description, characteristic for the developed period of the Metal Age. Metallurgic technology becomes a synthesis, which creates a completely new value in social relations raising them onto a higher level of abstraction. However, we may only speak of a complete synthesis only in the Bronze Age when metrological standards based on weight gain a greater spread in Europe (chapter 5).

The theoretical model mentioned above offers more space for additional speculations. After all it is not impossible that perceiving the proper rule of measuring metal occurred much earlier without being at present identified in archaeological material at the moment. While analyzing such discoveries as the beads from Colmar in chapter 5, I stated that their creator aimed at making specific portions of metal by the use of an inadequate method (stick manipulation). However, this method suggests at the same time that people were aware that a portion of metal is something more than its spatial measure–length. In those beads, the concept of length became not only a means of measurement in the sense of discovering a reference, which would be congruent with Kripke's interpretation (chapter 1), but a method for working out an objective and adequate value of metal. That is why the weight of beads dismembered in this way shows very precisely the real purpose. For the purpose was something which we would call a portion and similar behaviour had been noted in the case of Eneolithic Corded Ware and Bell Beaker vessels. Perhaps the concept of portion should be seen as a transitional phase in the forming of more abstract mathematical concepts.

Apart from the dialectic way of presenting the cognitive evolution described here we are able to outline its evolution. The basic idea was the measuring stick, embodied and burdened with a great deal of cosmological reference, which in the stage of initial development of metallurgy was transformed into the concept of portion, which in turn became connected with specific forms of metal objects. In the Bronze Age the form of that portion ceased to have any meaning as a full conceptualization of the weight of metal and its systematical weighing had been reached.

On this interpretational axis we may examine the metrologization of material culture. In the first stage the available quantification mechanisms were cognitive abilities, which we generally share with animals that are the quantification based on *subitizing*, a quick evaluation of the size/amount of objects. It was probably this mechanism that was responsible for the forming of three basic values in the first farming societies in Europe, as Pavlu presented it [76]. They were subject to evaluation, which could have been formulated as small, medium/large, and huge. In fact ceramic vessels do not require a more precise evaluation, although, as both research performed by me and by Georges show, at the end of the Neolithic period vessels could have been subject to mechanisms of rational, numerical evaluation, which was in general an exaggeration, an excess and probably a result of the culture specificity of the societies of that period[77]. From this perspective it would be easier to understand the peculiar behaviours in late Eneolithic societies, which were based on the ostentatious presence of measures and numbers. For these societies were the heirs of Eneolithic macrolithic traditions–a transformation of the measuring stick metaphor into its individualized form which was used for emphasizing social inequalities. That is why the measuring of vessels as part of the burial ritual should not raise our eyebrows (chapter 5) and because of this almost every burial was subject to this numerical characterization. This fixation with numbers/measures, emphasized by the seemingly egalitarian adornment of graves, does not remain without a connection with the earlier period when the concept of number/proto– number in human relations appeared.

---

[76] Pavlu 2000.
[77] Dzbynski 2004; Georges 2006, 609–613.

The Bronze Age–price, measure and the warrior's prestige

The next peak of the accumulation phenomenon falls during the Bronze Age, when social divisions developed strongly. According to researchers, their core was the chiefdom societies. The chief, who perhaps also fulfilled priestly functions, stood at the head of society. At lower positions in the hierarchy were the warriors subordinate to him, who did not have such functions. The chief controlled the flow of luxurious goods (metal objects, fanciful ornaments, etc.) while the whole structure was filled with "adrenaline and rivalry", numerous swords found and the signs of fighting which they bore being evidence of this[78].

In the Bronze Age we observe an increase of the consumption of prestige and wealth. Spheres of exchange developed dynamically, which connected areas distant from each other. Local manufacturing traditions were created, among which the Nordic culture is the best explored. Apart from developing local ornamentation characteristic bronze objects were made here: weapons, ornaments, tools and vessels. Great amounts of metal were included into rites and rituals at that time; there are numerous deposits of remains. Bronze axes, sickles, swords, bronze bars of different shapes etc had been deposited. The second phenomenon characteristic for that period were ostentatious tombs located under burrows equipped with great amounts of metal and other luxurious goods. Both swords as well as daggers were beautifully ornamented and adorned, which is a sign of high craftsmanship. Society was overtaken by the idea of fighting and celebrating is also reflected in stone engravings[79].

As a consequence, the transition from the Stone Age to the Bronze Age is marked with an ideological transformation which, according to Treherne, may be identified as a transition from the place and community ideology to the individual and personal expression ideology, which carried with it the adoption of an exceptionally ostentatious lifestyle. The new ideology of the Bronze Age was marked by putting an emphasis on social categorization, e.g. in the form of differentiating gender. Moreover, what we observe in the grave items is a "suspiciously luminous" stress, as Treherne writes, put on displaying the consumption of exotic and prestigious goods brought from far away countries[80].

The effective connection of tin, and copper went in unison with the development of specialized tools as well as swords, hence a fully characteristic male ethos appeared in the Bronze Age connected with war.

---

[78] Kristiansen 1984, 86.
[79] Kristiansen 1987, 30–51.
[80] Treherne 1995, 105–144; Shennan 1993, 121–161.

The connection of metal with its efficiency as well as high value had occurred, which caused a new form of identity in society. At the same time the acquired skill of horse riding adds up to this as well as the invention of the war chariot, and all these components create a very powerful network of meanings and metaphors, which determined most societies of the Metal Age in Europe, chivalry being its culmination in historic times.

Researchers see serious influences from the south of Europe in this process, especially from the Mycenaean culture, whose contents were so accurately described by Homer. The material culture of the Bronze Age seems to fit literary records, even though done long after its rise. However, we cannot forget the local conditionings for such a development, as a lot of evidence points to the fact that the evolutionary potential of the bronze society, in which the chief and warrior played a central role, is based deeply in the Eneolithic tradition of the beaker societies. Here the stress was put on the individual exhibition of status which demonstrated social inequality and did not hide it as was the case among earlier megalithic societies[81]. In this we may perceive one of the sources of the warrior aristocracy order.

The supplements of the warrior nature of the bronze societies are objects which to this day are treated somewhat as ornaments–massive bronze rings, which were most often made in the Carpathian and Alps regions. These rings were used as weight standards of metal symbolizing prestige and wealth. Some hoards, among which hundreds of such objects were found, are evidence of how such large amounts of objects changed place from one end of Europe to another.

According to Sherratt, in most cases the deposits should be treated as storages for travelling merchants or craftsmen[82]. He also noticed that the bronze exchange system was dominated by a very narrow spectrum of exchanged objects. Metal bars, although they had different forms, played the main role in this system which was the object of mass accumulation of metal. Part of this Bronze Age capital fell out of the global circulation from time to time and was deposited in hoards or in the tombs of the elite. Hence, the Bronze Age may be illustrated by two images: one of them presents impulsive men, overly taking care about their appearance, swinging swords and riding war chariots, while the other is the material base of the first–wealth measurable in metal, whose attribute is the bar, scale and weight[83].

---

[81] Thomas 1987, 421.
[82] Sherratt 1994, 244–76
[83] Pare 1999, 421–514; Peroni 1998, 217–224; Sommerfeld 1994.

The analyses of ring bars in Central Europe allow us to assume that they were made by following a highly abstract concept of the mass of the material. The major part of the bars analyzed by the German researcher Lenerz–De Wilde show maxima of the variable ranging from 180 grams to 200 grams. Moreover, she noted that with time new types of bars appeared which were much lighter. Their weight most often reaches approximately 90 grams. She interprets this phenomenon as being connected with the growing bar exchange "market", where "retail" transactions were perhaps performed, in which case smaller units of weight would have been be required.

Lenerz–De Wilde also considered the origin of this form of the bar in Europe, which was alien in its base[84]. Similar forms have been found in the deposit in Byblos, which have been dated between 2,130 BC and 2,040 BC. However, in time it appeared that many of the ring bars from Europe are older than the Levantine ones; hence the directions of the possible adaptations of this form had to be reversed. In addition, the existence of very similar forms from pure copper earlier on in the Eneolithic Baden culture in the area of present day Austria have been noted. Despite some hiatus between the early copper form and the later bronze bars, Lenerz–De Wilde allows for the idea that we are dealing in this case with a continuation of the ornamental ring form. This thesis is supplemented by the arguments formulated in this book. Not only is it ring bars that appear earliest in the area mentioned, but the so called axe bars appear also. Hence we may assume that at some point the connection of both ideas, which in effect resulted in a new, seemingly alien form of an artefact appeared, which was the effect of the growing abstractness of perception and the innovation of societies from the end of the Eneolithic period to the start of the Bronze Age.

The depositing of metal, the major part being the bars, was subject to certain cycles, which suggested some changes in ideology. In the early part of the Bronze Age accumulation is more distinct in the graves of wealthy warriors, while later on (in the middle of the Bronze Age) the mass depositing of great amounts of metal in the ground was popular. The items in graves became more symbolic at that time[85]. Weights also started to become a part of some graves. The practice of burying the dead changed from a skeletal burial to pyres. However, this does not explain the humble and rather symbolic set of items. Kristiansen guesses that this is connected with the stabilization of the social structure. In other words, when it was

---

[84] Lenerz–De Wilde 1995.
[85] Kristiansen 1991, 30.

not necessary to display the chiefs' wealth for the public, grave adornment became more symbolic. Hence, in the middle of the bronze period many graves contained bronze weights, which were used for the precise weighing of metal. The remaining grave items suggests that these graves belonged to people who had a high social status, although we cannot speak of specialists, such as metallurgists or craftsmen. Humble items in some of them, such as for example in Steinfurth, Germany, where apart from bronze pins only a set of weights had been found. The arrangement of the weights suggested that they were carried in some sort of pouch. Other burials are richer. In Richenmont–Pepinville, France, among the objects found there was a bronze sword, a razor, many other objects as well as some weights. Hence, the spectrum of grave adornment containing weights seems wide[86]. The first balance scales had appeared at that time, although weighing metal could have been performed earlier. Along with this, a change in the exchange system had occurred. Bronze sickles, which were also intensively fragmented, were deposited en mass instead of bars. The earlier exchange by the use of bars had been replaced by a system of the precise measuring of fragments of metal[87].

The research results made by archaeologists mentioned above clearly show that the processes of the "mathematization" of human relations, of measuring individual wealth in the Bronze Age were already far advanced. Some researchers think that comparable metrological systems functioned in most areas, the latest ones beginning from the middle of the epoch, as part of the framework of the "weighed money" economy[88]. This type of economy is characteristic for civilization peripherals while the centre was composed of the Aegean palace civilization. Regardless of the detailed interpretation, the Bronze Age material culture clearly shows that the major part of these societies were able to measure/count metal in an abstract form, that is treating it as weight. The measuring stick idea was not necessary to evaluate the amount of metal from a long time which, although it became a luxurious product, it was also a mass product.

The social context of the evolution of the measuring stick metaphor

We already know that we can define the cultural transformation in Europe as a transition from the idea of place and community to the idea of

---

[86] Pare 1999.
[87] Sommerfeld 1994.
[88] Peroni 1998.

individual expression[89]. However, we may also say that it was a transformation based on a transition from stone technology to metal technology.

Kristiansen proposed[90], that most of the Neolithic age should be seen as a transformation but also as a discourse between two systems of social organization. The use of monumentality and burial rituals was characteristic for the first one, which was the product of social structures dubbed territorial chiefdoms. The second system of social organization, which came late as far as chronology is concerned, were segmentary societies, roughly identified with the beaker cultures. Megalithic societies were characterized by an eminent farming economy, burning forests for farming and probably also for building megaliths. Such a type of farming, visible in pollen charts, had been confirmed in locations very distant from each other, such as Scandinavia and south east Poland[91]. The settling of that period seems to acquire features of a central settlement model (although developed differently depending on the region), tendencies towards establishing entrenched settlements or enclosures of a ritual character. This settling shows a considerable complexity which points to the great meaning of the settlement or place of community meetings, which I described in chapter three. Another characteristic element of this socio–cultural system described here became the production of macrolithic tools: flint axes and blades, which had both a practical as well as a symbolic meaning[92]. This phenomenon was the subject of different fluctuations. Kristiansen points out to the fact that e.g. in the later phases of megalithic societies the basic objects of social interest–axes seem to be connected more with graves than with hoards. The essence of this trend seems to be the Globular Amphorae societies, which I described in chapter four.

As we remember from chapter three, the earliest megalithic constructions were the long tombs, often constructed from wood and earth, which refer to the long houses of the early farmers. In the subsequent phase new varied forms appeared which was evidence of a progressing social discourse, whose content, as researcher's state, was probably connected with the changing relationships between the authorities of specific families or lineages[93]. The aim of megalithic founding was to emphasize the social position of a small group of people, although their

---

[89] Treherne 1995; Renfrew 2007, 168.
[90] Kristiansen 1994.
[91] Milisauskas & Kruk 1999, 147; 1989, 403–446.
[92] Kristiansen 1984, 78.
[93] Müller 2001, 424.

communication and solidarity aspect, which rather emphasizes the social anonymity, is also clear[94]. Subsequent deceased's were rarely added, celebrations were held near the tombs, further grave gifts were offered, which do not, however, allow themselves to be associated with specific individuals (the Globular Amphorae society being sometimes an exception to this rule).

The later beaker traditions, whose characteristic features are single graves covered with a barrow, which according to Kristiansen has a distant connection to the megalithic tradition, stands in complete opposition to the above culture model. However, here the individuality and its attributes, connected with gender and social rank, took centre stage of social attention. A standard set of objects was placed inside a grave which characterized the social rank, age and gender of the deceased. Moreover, among the beaker cultures there is very limited evidence of rituals connected with depositing valuable objects and communal rituality in general. This seemed a complete break from the former tradition, although, as Charlotte Damm points out, it refers to that tradition by means of contrast and discourse[95]. Kristiansen approaches this matter similarly. According to him megalithic culture connects the economy, social organization and religion into one coherent system of a ritualized sphere of reproduction, whereas beaker societies seem to separate this institution, being dominated by a horizontal, competitive structure which stood opposite to the community that was based on a local economical autonomy with a faint share (if any) of communal ritruality. Earlier societies left varied and rich evidence of farming, which is also evidence of social organization and religion. Whereas the later Eneolithic societies left behind a poor and narrow material spectrum found mainly in the graves. Hence, according to Kristiansen we would be faced here with (early) farming *versus* (later) a pastoral economy[96].

On the other hand, Julian Thomas underlined that in many parts of central and northern Europe between the second half of the 4th millennium BC and the end of the 3rd millennium BC the communal grave and ritual assumptions coexisted with individual burials[97]. Studies performed by Müller also show the growing complexity of tomb constructions in a scale of time, whereas finally somewhere in the first half of the 3rd millennium BC individual burial became the dominating form[98]. Again, an excellent

---

[94] Damm 1991, 143.
[95] Ibidem.
[96] Kristiansen 1991, 85.
[97] Thomas 1987, 422.
[98] Müller 2001.

example of the described development is the Globular Amphorae societies, among which both trends had their place. While not being antagonistic (group or individual) and where, moreover, the role of macrolithic axes, as objects which probably had a numerical meaning, is clearly marked. I wrote of this in chapter four.

In the socio–economical transformation path outlined above we should also place cognitive transformations connected with the development of mathematical perception, concepts of number as well as more rational systems of value. The first stage would be, consequently, connected with the discourse constituted with varied uses of the measuring stick metaphor in a wide spectrum of communal behaviours. This is seen both in settlements in the south of Europe as well as in long houses in the north. Leading out of the measuring stick metaphor itself beyond settling ideas, and its introduction into pure cult ideas, is a sign of a serious transformation, of which many researchers are perfectly aware. Neolithic settlements and houses, which in the beginning were the centre of social interaction, became replaced in two ways: on the one hand by a dynamically developing ritual sphere, still connected with the community but led out beyond the settlements, and on the other hand by a sphere of individualized communication, during which–in the framework of technological development–the concept of number/measure was subject of propagation and intensification. Hence, in the Eneolithic period the area of discourse grew significantly. On the one hand, the metaphorization of the line/stick became harnessed to realize the ideas of a cultural and social character, such as megaliths, enclosures, etc.; while on the other hand, it became the subject of the process of individualization in the form of macrolithic industries.

However, this was not the process of a universal dimension. There occurred a structural crack between the northern and southern areas of Central Europe which determined the further development of societies. In my opinion, we are faced here with another fundamental "cave abandoning" type of division. As architecture bonded societies into a more coherent oriented community communication, its near total abandonment in the southern regions of Europe, which was the separation of graves from the settlements and establishing flat cemeteries which forced the creation of new symbols that did not appear out of nowhere. This is what we most often have in mind when speaking about transformation. Macrolithic blades, which were a transformation of the measuring stick metaphor, became this symbol. However, architecture in the north bonded human minds stronger and for a longer period, first in the form of houses, next of megalithic constructions and finally of monumental constructions.

That is why the development of macrolithic industries in this region must be examined in this mental background. The individualized outline of macrolithic industries clashed for a longer time with the communal version of the stick metaphor that was present in architecture in the northern regions of Europe, while in the south it quickly became an element of intersubjective relations reflected in the graves from that period. **In other words, similarly as architectural constructions replaced caves at the end of the Palaeolithic period, so macroliths replaced architectural constructions in the Eneolithic, thus creating a new symbolic frame to be filled, inter alia with mathematical concepts.**

However, if we are discussing graves, and not living beings, may we consider this image as a valuable illustration of the actual relations between the living? Not fully, of course, although we cannot assume that studies concerning the burial ritual are a world which is cognitively unapproachable for us and which has no connection with the world of people living in those times. It was Binford's analyses which showed that the burial rituals of simple societies reflect the views of society and the world (gender, age, wealth, etc.)[99]. In our case it is enough to remember that the burial ritual in the Neolithic and Eneolithic periods was so strongly connected with the place, house and settlement symbolics, that it was something similar to a displaced area of symbolic activity. The disposing of ancestors from the settlement area had to leave some traces of communal awareness. In connection to this, have we not enough reasons to state that the grave items being symbolic is another transformation of the *domus* area, during which new means of symbolic expression appeared?

It was probably the early Neolithic societies which developed a humble, simple base of metrological–numerical concepts which were shaped as a result of the manipulation of the measuring stick metaphor in body–architecture relations. It was more of a type of new grammar and vocabulary for defining basic relations: a whole, a half, a quarter and three quarters. This grammar was subsequently implemented into the sphere of tool production. The cognitive bonding function in this transformation was still performed by the measuring stick metaphor. At this stage the objects of social discourse which bore the potential for revolutionary changes were axes and macrolithic flint blades. The most important element for this revolution was their potential for the individualization of human behaviour through assigning metrological features to these objects, that is numerical values, which is reflected in their varied fragmentation.

---

[99] Binford 1971, 6–29.

The next stage, connected with the beaker societies, was characterized by the development of a new description of reality by means of more and more abstract mathematical concepts. It was probably in this period that the concept of an abstract weight was introduced for the first time, which the discovery of axe bars and copper beads from Kelsterbach point to, or it was on its path to complete conceptualization. Both my earlier research concerning the capacity of Corded Ware vessels as well as George's research[100] point to the fact that the transformations of the measuring stick were at a very advanced stage, for which the best definition at present is the concept of portions. Both the vessels as well as metal objects could have been manufactured by coiling the stick, but their final form did not resemble it at all. In other words, the number/measure concepts introduced into social structures in the Eneolithic period found their consequence finally in societies living at the end of the period, where gender, age and individual status were defined numerically, which is a sort of sensation.

When summarizing this development of prehistoric European communities from the Neolithic period to the developed Metal Age we have to finally underline the greatest transformation. In the beginning we are faced with the subjugation of groups and individuals to monumental linearity in different forms, thus creating a sphere of socio–economical activity. Both the living and the dead were built into a strict architectural frame obedient to the cosmic laws and seen as a fragment of these structures, almost equal with the objects which seem to circle around them, e.g. as part of the relation "enchainment" phenomenon. Only a few were granted the honour of directly experiencing contact with the manifestation of the line connecting both worlds. The distance towards objects/things was clearly marked, which is reflected in the basic character of early exchange systems. In truth at that time objects did not belong to anyone as they circulated in a sphere of timeless narration.

At the end of the Eneolithic period the human body definitely takes on the role of the reference line. Monumental linearity finally became internalized in the body, which became the object of fascination[101]. Now the dead were being oriented in astronomic directions without the mediation of architectural structures, whereas they became surrounded by objects according to strict rules, which both define each deceased as well as express the social attitude towards objects as being close to man as their property. It is here we first find a rather clear concept of man being the measure of all things. And so it was in reality. A reversal of symbolic

---

[100] Georges 2006.
[101] Treherne 1995.

order, which was the context of earlier beliefs, took place, the process of the metrologization of the material world–a cognitive base for later mathematization–being a significant factor in this process. Undoubtedly, this transformation is to be perceived as a consequence of development, during which man realized that he could enforce his ideas on nature. The measuring stick metaphor was finally internalized, transformed and acknowledged as a human creation–weight. And is it not so that measurement throws the whole world before man?[102]

---

[102] In the original: "Die Kunst des Messens unterwirft dem Menschen die Welt; durch die Kunst des Schreibens hört seine Erkenntnis auf, so vergänglich zu sein, wie er selbst ist; sie beide geben dem Menschen, was die Natur ihm versagte, Allmacht und Ewigkeit". Theodor Mommsen, Römische Geschichte, erstes Buch.

BIBLIOGRAPHY

Aitchison 2000: J. Aitchison, The Seeds of Speech: Language Origin and Evolution. Cambridge: Cambridge University Press.
—. 2000a: Ziarna mowy. Warszawa: PIW.
Anderson 1958: W. F. Anderson, Arithmetical Procedure in Minoan Linear A and in Minoan–Greek Linear B. American Journal of Archaeology, Vol. 62, No. 3: 363–368.
Aveni 1982: A. F. Aveni, Archaeoastronomy in the New World. Cambridge: Cambridge University Press.
Bácskay & Siman 1987: E. Bácskay, K. Simán, Some remarks on chipped stone industries of the earliest Neolithic populations in present Hungary. In: Chipped Stone Industries of the Early Farming Cultures in Europe. Archaeologia Interregionalis, 107–130. Warszawa: Wydawnictwa Uniwersytetu Warszawskiego.
Bailey 1990: D. W. Bailey, The Living House: Signifying Continuity. In: R. Samson (ed.), The Social Archaeology of Houses: 19–48. Edinburgh: Edinburgh University Press.
—. 2000: D. W. Bailey, Balkan Prehistory. Exclusion, incorporation and identity. London–New York: Routledge.
Bakels 1987: C. C. Bakels, On the Adzes of the Northwestern Linearbandkeramik. Analecta Praehistorica Leidensia 20: 53–85.
Balcer 1983: B. Balcer, Wytwórczość krzemienna w neolicie ziem polskich. Warszawa: Zakład Narodowy im. Ossolińskich.
—. 2002: B. Balcer, Ćmielów, Krzemionki, Świeciechów. Związki osady neolitycznej z kopalniami krzemienia. Warszawa: Instytut Archeologii i Etnologii PAN.
Barkow 1975: J. Barkow, Prestige and Culture. A Biosocial Interpretation. Current Anthropology, Vol. 16, No. 4: 553–572.
Barrow 1996: J. Barrow, Warum die Welt mathematisch ist. München: Deutscher Taschenbuch Verlag GmbH.
—. 1999: J. Barrow, Ein Himmel voller Zahlen. Auf den Spuren mathematischer Wahrheit. Reinbeck: Rowohlt Taschenbuch Verlag GmbH.
Bateson 1996: G. Bateson, Umysł i przyroda. Warszawa: PIW.

Becker 2008: V. Becker, Early and middle Neolithic figurines–the migration of religious belief. Documenta Praehistorica XXXIV: 119–127.
Behn 1938: F. Behn, Ein Grabfund der Steinkupferzeit von Kelsterbach, Starkenburg. Germania 22: 77–78.
Bellwood 2005: P. Bellwood, First Farmers. The Origins of Agricultural Societies. Oxford: Blackwell.
Benedict 1999: R. Benedict, Wzory kultury. Warszawa: MUZA.
Besse 2004: M. Besse, Bell Beaker common ware during the third Millenium BC in Europe. In: J. Czebreszuk (ed.), Similar but Different. Bell Beakers in Europe, 127–148. Poznań: Adam Mickiewicz University.
Biehl 2003: P. Biehl, Studien zum Symbolgut des Neolithikums und der Kupferzeit in Südosteuropa, Saarbrücker Beiträge zur Altertumskunde 64. Bonn: Habelt Verlag.
Binford 1971: L. Binford, Mortuary practices: their study and their potential. In: J. A. Brown (ed.), Approaches to the Social Dimensions of Mortuary Practices. Memoirs of the Society for American Archaeology, No. 25: 6–29.
Bognar–Kutzian 1963: I. Bognar–Kutzian, The Copper Age Cemetery of Tiszapolgar–Basatanya. Archaeologica Hungarica 42. Budapest: Akademiai Kiado.
Boric 2008: D. Boric, First Households and „House Societies" in Europaean Prehistory. In: A. Jones (ed.), Prehistoric Europe. Theory and Practice: 109–142. Oxford: Blackwell.
Borkowski & Migal 1996: W. Borkowski, W. Migal, Ze studiów nad użytkowaniem siekier czworościennych z krzemienia pasiastego. Studia nad gospodarką surowcami krzemiennymi w pradziejach t. 3. Warszawa: Państwowe Muzeum Archeologiczne.
Borkowski 1995: W. Borkowski, Krzemionki Mining Complex. Studia nad gospodarką surowcami krzemiennymi w pradziejach, t. 2. Warszawa: Państwowe Muzeum Archeologiczne.
Bourdieu 1973: P. Bourdieu, The Berber house. In: M. Douglas (ed.), Rules and Meaning. Harmondsworth: Penguin.
Bradley 1998: R. Bradley, The Significance of Monuments. On the shaping of human experience in Neolithic and Bronze Age Europe. London–New York: Routledge.
—. 2001: R. Bradley, Orientations and origins: a symbolic dimension to the longhouse in Neolithic Europe. Antiquity 75: 50–56.
Broadbent 1955: S. R. Broadbent, Quantum Hypotheses. Biometrika 42/1: 45–57.

Brumfiel & Earle 1987: E. Brumfiel, T. Earle, Specialization, exchange and complex societies: an introduction. In: Brumfiel E, Earle T. (eds.), Specialization, Exchange and complex societies: 1–9. Cambridge: Cambridge University Press.
Budja 1999: M. Budja, Clay tokens–accounting before writing in Eurasia. Documenta Praehistorica XXV: 219–235.
—. 2003: M. Budja, Seals, contracts and tokens in the Balkans Early Neolithic: where in the puzzle. Documenta Praehistorica XXX: 115–130.
Budziszewski & Tunia 2000: J. Budziszewski, K. Tunia, A Grave of the Corded Ware Culture Arrowheads Producer in Koniusza, Southern Poland. Revisited. In: Kadrow S. (ed.), A Turning of Ages. Jubilee Book Dedicated to Professor Jan Machnik on His 70th Anniversary: 101–136. Kraków: Institute of Archaeology and Ethnology Polish Academy of Sciences.
Budziszewski 2000: J. Budziszewski, Flint working of the south–eastern group of the funnel beaker culture: exemplary Deception of chalcolithic socio–economic patterns of the pontic zone. Baltic–Pontic Studies 9: 256–281.
—. 2006: J. Budziszewski, Flint Economy in Chalcolithic Societes of East–Central Europe, Veröffentlichungen aus dem Deutschen Bergbau–Museum Bochum 148: 315–321.
Case 1995: H. Case, Beakers: loosening a stereotype. In: I. Kinnes, G. Varndell (eds.), Unbaked Urns of Rudely Shape: 55–67. Oxford: Oxbow Monograph 55.
Cauvin 2000: J. Cauvin, The birth of the Goods and the origin of Agriculture. Cambridge: Cambridge University Press.
Chapman 1990: J. Chapman, Social Inequality on Bulgarian Tells and the Varna Problem. In: R. Samson (ed.). The Social Archaeology of Houses: 49–92. Edinburgh: Edinburgh University Press.
—. 2000: J. Chapman, Fragmentation in Archaeology. People, places and broken objects in the prehistory in south–eastern Europe. London: Routledge.
Chapman et al. 2008: J. Chapman, B. Gaydarska, V. Slavchev, The life histories of Spondylus shell rings from the Varna I Eneolithic cemetery (Northeast Bulgaria): transformation, revelation, fragmentation and deposition. Acta Musei Varnaensis VI: 139–162.
Clottes & Lewis–Williams 1996: J. Clottes, D. Lewis–Williams, Les chamanes de la préhistoire. Text intégral, polémique et réponses. Paris.

Clottes & Lewis–Williams 2009: J. Clottes, D. Lewis–Williams, Prehistoryczni szamani. Trans i magia w zdobionych grotach. Warszawa: Wydawnictwa Uniwersytetu Warszawskiego.

Coudart 1993: A. Coudart, De l'usage de l'architecture domestique et de l'anthropologie sociale dans l'approche des societes neolithiques: 'exemple du neolithique danubien. Documents d'archeologie francaise n 41: 114–135.

Cvek 2004: O. V. Cvek, Goncharnye vyrobnitstvo plemen trypilskoy kultury. In: M. Videiko, N. Burdo (eds.), Encyclopedia of the Trypillian Civilization in two volumes. Kiev.

Czerniak 1998: L. Czerniak, The First Farmers. In: Pipeline of Archaeological Treasures: 23–36. Poznań: Poznańskie Towarzystwo Prehistoryczne.

Damasio 1994: A. Damasio, Descartes' Error: Emotion, Reason, and the Human Brain. New York: Putnam Publishing.

—. 1999: A. Damasio, Błąd Kartezjusza. Poznań: Rebis 1999.

Damerow & Englund 1987: P. Damerow, Englund R.K., Die Zahlzeichensysteme der Archaischen Texte aus Uruk. In: (red.) M.W. Green, H.J. Nissen, Zeichenliste der Archaischen Texte aus Uruk: 117–166. Berlin.

Damerow 1996: P. Damerow, Prehistory and Cognitive Development. Max–Planck–Institut for the History of Science, Preprint 30, http://www.mpiwg–berlin.mpg.de/Preprints/P30.PDF

—. 1999: P. Damerow, The Material Culture of Calculation. A Conceptual Framework for an Historical Epistemology of the Concept of Number. Max–Planck–Inst. Für Wissenschaftsgeschichte, Preprint 117, http://www.mpiwgberlin.mpg.de/Preprints/P117.PDF

Damm 1991: C.B. Damm, Continuity and Change. An Analysis of Social and Material Patterns in the Danish Neolithic. Cambridge: Cambridge University Press.

Davies 2012: S. R. Davies, The Early Neolithic Tor Enclosures of Southwest Britain, http://etheses.bham.ac.uk/1141/1/Davies_S_10_Ph D.pdf

—. 1983: A. Davis, Pacing the megalithic yard. Glasgow Archaeological Journal. Volume 10: 7–11.

Diakonof 1983: I. M. Diakonof, Some Reflections on Numerals in Sumerian towards a History of Mathematical Speculations. Journal of the American Oriental Society 103(1): 83–93.

Dohrn–Ihmig 1983: M. Dohrn–Ihmig, Das bandkeramische Gräberfeld von Aldenhoven–Niedermerz, Kreis Düren, Rheinische Ausgrabungen, Band 24: 47–179.

Durman 2001: A. Durman, Celestial symbolism in the Vučedol culture. Documenta Praehistorica XXVIII: 215–226.
Dzbynski & Wiermann 2002: A. Dzbyński, R. Wiermann, Von Alten, Äxten und Amphoren. Praehistorica XXV/XXVI: 205–226.
Dzbynski 2001: A. Dzbyński, Die Schnurkeramik in Böhmen und ihre Gefäßvolumina. Basel (unpublished manuscript).
—. 2004: A. Dzbyński, Metrologische Strukturen in der Kultur mit Schnurkeramik und ihre Bedeutung für die Kulturentwicklung des mitteleuropäischen Raumes. Beiträge zur Ur–und Frühgeschichte Mitteleuropas 39. Langenweissbach: Beier, Beran Verlag.
—. 2008: A. Dzbyński, Rytuał i porozumienie. Racjonalne podstawy komunikacji i wymiany w pradziejach Europy Środkowej. [Ritual and Understanding. Rational Bases of Communication and Exchange in Prahistoric Central Europe]. Rzeszów: Mitel.
—. 2008a: A. Dzbyński, Von Seeberg bis Kelsterbach: ein Beitrag zur Bedeutung des Kupfers im Äneolithikum und in der Bronzezeit Europas. Prähistorische Zeitschrift, Band 83: 36–44.
—. 2010: A. Dzbyński, Das Häuptlingscourt von Vikletice–die soziale Differenzierung in der Schnurkeramik. In: E. Claßen, T. Doppler, B. Ramminger (Hrsg.), Familie–Verwandtschaft–Sozialstrukturen: Sozialarchäologische Forschungen zu neolithischenn Befunden. Fokus Jungsteinzeit. Berichte der AG–Neolithikum 1. Kerpen–Loogh 2010: 179–183.
—. 2011: A. Dzbyński, Mr. Blademan. Macrolithic technology — Eneolithic vocabulary and metaphors. Documenta Praehistorica XXXVIII, Neolithic Studies 18: 172–184.
—. 2011a: A. Dzbyński, Pan Wiórecki i Świat–Maszyna. Poznań: SORUS.
Dzieduszycka–Machnikowa 1961: A. Dzieduszycka–Machnikowa, Z zagadnień krzemieniarstwa neolitycznego. Sprawozdania z posiedzeń Komisji Naukowych Oddziału PAN w Krakowie, Kraków styczeń–czerwiec 1961: 29–31.
Eliade 1993: M. Eliade, Kowale i alchemicy. Warszawa: Aletheia.
Engels 1884: F. Engels, Der Unsprung der Faimilie, des Privateigenthums und des Staats Im Anschluss an Lewis H. Morgans Forschungen. Stuttgart.
Everett & Madora 2012: C. Everett, K. Madora, Quantity Recognition Among Speakers of an Anumeric Language. Cognitive Science 36: 130–141.
Frank et al. 2008: M. Frank, D. Everett, E. Fedorenko, E. Gibson, Number as a cognitive technology: Evidence from Pirahã language and cognition. Cognition 108: 819–824.

Frolov 1974: B.A. Frolov, Čisla w grafike Paleolita. Novosibirsk: Nauka, Sib. otd–nie.

Gamble 2004: C. Gamble, Materiality and Symbolic Force: a Paleolithic View of Sedentism. In: E. DeMarrais, Ch. Gosden, C. Renfrew (eds.), Rethinking materiality. The engangement of mind with the material world, 85–95. Cambridge: McDonald Institute of Archaeological Research.

Gates 2003: Ch. Gates, The Archaeology of Urban Life in the Ancient Near East and Egypt, Grece and Rome. London–New York: Routledge.

Georges 2006: V. Georges, La volumétrie dans les agrosystèmes préhistoriques: céramique étalon (campaniforme) ou instrument de mesure comlexe? Le contenant céramique et le modèle campaniforme. Bulletin de la Société préhistorique française 103/3: 609–613.

Geslin et al. 1980: M. Geslin, G. Bastien, N. Mallet, Das Klingendepot von La Creusette, Gem. Barrou, Dep. Indre, Lore. In: 5000 Jahre Feuersteinbergbau. Veröffentlichungen aus dem Deutschen Bergbau–Museum Bochum, Nr 22: 289–293.

Gordon 2004: P. Gordon, Numerical cognition without words: Evidence from Amazonia. Science 306:496–499.

Gronenborn 2003: D. Gronenborg, Der „Jäger/Krieger" aus Schwanfeld. Einige Aspekte der politisch–sozialen Geschichte des mitteleuropäischen Altneolithikums. In: J. Eckert, U. Eisenhauer, A. Zimmermann(Hrsg.), Archäologische Perspektiven. Analysen und Interpretationen im Wandel: 35–48. Rahden/Westf.: Verlag Marie Leidorf GmbH.

Gruber & Voneche 1977: H. E. Gruber, J. J. Voneche (eds.), The Essential Piaget. London: Routledge.

Haarmann 2006: H. Haarmann, The challenge of the abstract mind: symbols, signs and notation systems in Europaean prehistory. Documenta Praehistorica XXXII: 221–232.

Habermas 1981: J. Habermas, Theorie des kommunikativen Handelns. (Bd.1: Handlungsrationalität und gesellschaftliche Rationalisierung), Frankfurt am Main: Suhrkamp.

—. 1999: J. Habermas, Teoria działania komunikacyjnego. Tom I: Racjonalność działania a racjonalność społeczna. Warszawa: PWN.

—. 2002: J. Habermas, Teoria działania komunikacyjnego. Tom II: Przyczynek do krytyki rozumu funkcjonalnego. Warszawa: PWN.

—. 2003: J. Habermas, Przyszłość natury ludzkiej. Warszawa: Scholar.

Hahn 1982: J. Hahn, Eine menschliche Halbreliefdarstellung aus der Geißenklösterle–Höhle bei Blaubeuren. Fundberichte Baden–Württemberg 7: 1–12.

Hansen 2001: S. Hansen, Fruchtbarkeit? Zur Interpretation neolithischer und chalkolithischer Figuralplastik, Mitteilungen der Anthropologischen Gesellschaft in Wien, Band 130/131: 93–106.

Häusler 1981: A. Häusler, Zu den Beziehungen zwischen dem Nordponitischen Gebiet, Südost und Mitteleuropa im Neolithikum und in der frühen Bronzezeit und ihre Bedeutung für das indoeuropäische Problem. Przegląd Archeologiczny Vol. 29: 101–149.

—. 2007: A. Häusler, Zu den Bestattungssitten der Tripol'e–Kultur und der neolithischen–äneolithischen Kulturen Südosteuropas. In: M. Stefanovich, Chr. Angelova (eds.), PRAE. In Honorem Henrieta Todorova: 55–77.

Havel 1978: J. Havel, Pohrebni ritus kultury zvoncovitych pucharu v Cechach a na Morave [The Burial Rite of the Bell Beaker Culture in Bohemia and Moravia]. Praehistorica VII: 91–117.

Heidegger 1982: M. Heidegger, The Question Concerning Technology, and Other Essays. New York: Harper TorchBooks.

Hemmy 1938: A.S. Hemmy, System of Weights at Mohenjo–Daro. In: E.J.H. Mackay (ed.), Further Excavations at Mohenjo–Daro: 601–612. Delhi: Government of India Press.

Hodder 1990: I. Hodder, The Domestication of Europe. Oxford: Blackwell.

—. 2004: I. Hodder, Neo–thingness. In: J. Cherry, C. Scarre, S. Shennman (eds.), Explaining social change: studies in honour of Colin Renfrew: 45–52. Cabridge: McDonalds Institute Press.

—. 2006: I. Hodder, The Leopard's Tale: Revealing the Mysteries of Catalhoyuk. London: Thames, Hudson.

Ifrah 1985: G. Ifrah, From one to zero. A universal history of numbers. New York: Viking Penguin.

—. 2006: G. Ifrah, Historia powszechna cyfr. T. 1. Warszawa: Wyd. W.A.B.

Ivanova 2008: M. Ivanova, Befestigte Siedlungen auf dem Balkan, in der Aegais und in Westanatolien ca. 5000–2000 v. Chr. New York–Berlin: Waxmann Verlag.

—. 2007: V. Vs. Ivanov, Towards Semiotics of Number. Bulletin of the Georgian National Academy of Science, N. 175: 186–194.

Jansen 2010: M. Jansen 2010, Architectural measurements in the Indus cities: The case study of Mohenjo–Daro. In: I. Morley, C. Renfrew (eds.), The Archaeology of Measurement. Comprehending Heaven, Earth and Time in Ancient Societies: 125–129. Cambridge: Cambridge University Press.

Jensen 2000: O.W. Jensen 2000, Between body and artefacts. Merleu–Ponty and archaeology. In: Holtorf, Karlsson (ed.), Philosophy and Archaeological Practice. Perspectives for 21th Century. Goeteborg.

Justenson 2010: J. Justenson, Numerical cognition and the development of 'zero' in Mesoamerica. In: I. Morley, C. Renfrew (eds.), The Archaeology of Measurement. Comprehending Heaven, Earth and Time in Ancient Societies: 43–53. Cambridge: Cambridge University Press.

Justus 1999: C. F. Justus, Pre–decimal structures in counting and metrology. In: (red.) E.C. Polome, C.F. Justus, Language change and typological variation: 55–79. Washington.

Kaczanowska & Kozłowski 2008: M. Kaczanowska, J. K. Kozłowski, The Körös and the Early Eastern Linear Culture in the Northern Part of the Carpathian Basin: A View from the Perspective of Lithic Industries. Acta Terrae Septemcastrensis 7: 9–37.

Kaflińska 2006: M. Kaflińska, Neolityczne depozyty gromadne na ziemiach polskich. Materiały i sprawozdania Rzeszowskiego Ośrodka Archeologicznego XXVII: 5–26.

Kahlke 2004: H–D. Kahlke, Sonderhausen und Bruchstedt. Zwei Gräberfelder mit älterer Linienbandkeramik in Tühringen. Weimarer Monographien zur Ur–und Frühgeschichte, Band 39.

Karlovsky & Pavuk 2002: V. Karlovsky, J. Pavuk, Analyza rozmerov domov lengyelskej kultury [Dimension analysis of Lengyel cuture houses]. Archeologicke Rozhledy LIV: 137–156.

Kendall 1974: D. G. Kendall, Hunting Quanta. Phil. Trans. Roy. Soc. London 276: 231–266.

Kibbert 1980: B.K. Kibbert, Die Äxte und Beile im mittleren Westdeutschland I. Prähistorische Bronzefunde IX, 10. München.

Klassen 2001: L. Klassen, Frühes Kupfer im Norden. Untersuchungen zu Chronologie, Herkunft und Bedeutung der Kupferfunde der Nordgruppe der Trichterbecherkultur. Jutland Archaeological Society, vol. 36.

Koch 1998: E. Koch, Neolithic Bog Pots from Zeeland, Mon, Lolland and Falster. Det Kongelige Nordiske Oldskriftselskab. Kobenhavn.

Koestler 1968: A. Koestler, The Sleepwalkers. New York: Macmillan Press.

Kołakowski 2009: L. Kołakowski, Główne nurty marksizmu. Warszawa: PWN.

Kordys 2006: J. Kordys, Kategorie antropologiczne i tożsamość narracyjna. Szkice z pogranicza neurosemiotyki i historii kultury. Kraków: Universitas.

Kovarova 2004: T. Kovarova, The spatial distribution of artefacts in Corded Ware graves. In: L. Smejda, J. Turek (eds.), Spatial Analysis of Funerary Areas. Plzeň: 21–37.
Kowalski 2000: A. Kowalski, Najstarsza metalurgia. Od sensów magicznych do wartości estetycznych. In: H. van den Boom, A.P. Kowalski, M. Kwapiński (red.), Kultura archaiczna w świetle wyobrażeń, słów i rzeczy: 195–215. Gdański: Muzeum Archeologiczne.
Kozłowski & Kaczanowska 1990: J. K. Kozłowski, M. Kaczanowska, Chipped Stone Industry of the Vinca Culture. In: Srejović, D. and N. Tasić (eds.), Vinca and its World. International Symposium The Danubian Region from 6000 to 3000 B.C., Beograd. Serbian Academy of Sciences and Arts, Centre for Archaeological Research of the Faculty of Philosophy: 35–48.
Krause 2001: R. Krause, Grabenwerk–Siedlung–Graeberfeld: Die Ausgrabungen von 1994–1997 von Vaihingen an der Enz (Kr. Ludwigsburg, Baden–Württemberg), Preistoria Alpina 37: 109–136.
Kripke 2001: S. Kripke, Nazywanie a konieczność. Warszawa: Aletheia.
Kristiansen 1984: K. Kristiansen, Ideology and material culture: an archaeological perspective. In: M. Spriggs (ed), Marxist Perspectives in Archaeology: 72–100. Cambridge: Cambridge University Press.
—. 1987: K. Kristiansen, From Stone to Bronze: The Evolution of Social Complexity in Northern Europe. In: (red.) Brumfiel E, Earle T., Specialization, Exchange and complex societies: 30–51. Cambridge: Cambridge University Press.
—. 1991: K. Kristiansen, Chiefdoms, states, and systems of social evolution. In: Earle T. (ed.), Chiefdoms: Power, Economy and Ideology: 16–43. Cambridge: Cambridge University Press.
Krištuf 2005: P. Krištuf, Džbány českého eneolitu (Jugs in the Eneolithic in Bohemia). In: E. Neustupný, J. John (eds.), Příspěvky k archeologii 2, Plzeň: KAR FF ZČU: 69–118.
Kruk & Milisauskas 1989: J. Kruk, S. Milisauskas, Neolithic economy in central Europe. Journal of World Archaeology t. 3, nr 4: 77–96.
Kruk & Milisauskas 1999: J. Kruk, S. Milisauskas, Rozkwit i upadek społeczeństw rolniczych neolitu [The Rise and Fall of Neolithic Societies]. Kraków: Instytut Archeologii i Etnologii PAN.
Krzak 1994: Z. Krzak, Megality Europy. Warszawa: PWN.
Kubba 1998: S.A.A. Kubba, Architecture and Linear Measurement during the Ubaid Period in Mesopotamia. BAR International Series 707.
Kuckenburg 2004: M. Kuckenburg, Wer sprach das erste Wort. Die Entstehung von Sprache und Schrift. Stuttgart: Theiss Verlag.

Kula 1986: W. Kula, Measures and Man. Princeton: Princeton University Press.
—. 2004: W. Kula, Miary i ludzie. Warszawa: PWN.
Lakoff & Johnson 2003: G. Lakoff, M. Johnson, Metaphors We Live By. Chicago: The University of Chicago Press.
Lakoff & Núñez 2000: G. Lakoff, R. Núñez, Where Mathematics Comes From. How the Embodied Mind Brings Mathematics into Being. New York: Basic Books.
Last 1996: J. Last, Neolithic Houses–A central Europaean perspective. In: T. Darvill, J. Thomas (eds.), Neolithic Houses in Northwest Europe and Beyond: 27–40. Oxford: Oxbow Books.
Latour 1993: B. Latour, We Have Never Been Modern. Harvard University Press.
Leach 1976: E. Leach, Culture and communication. The logic by which symbols are connected. Cambridge: Cambridge University Press.
—. 1991: J. Lech, The Neolithic–Eneolithic transition in prehistoric mining and siliceou rock distribution. In: J. Lichardus (ed.), Die Kupferzeit als historiche Epoche: 557–574. Symposium Saarbrücken und Otzenhausen 6–13.11.1988. Bonn.
Lee 1978: R. Lee, Polithisc, sexual and non–sexual in an egalitarian society. Social Science Information 17 (6): 871–895.
Lefranc et al. 2009: P. Lefranc et al., Inhumations, dépôts d'animaux et perles en cuivre sur le site néolithique récent de Colmar "aérodrome". In : 10 000 ans d'histoire ! Dix ans de fouilles archéologiques en Alsace, Musées de la ville de Strasbourg. Catalogue du Musée archéologique de Strasbourg: 43–45.
Lenerz De–Wilde 1995: M. Lenerz De–Wilde, Prämonäre Zahlungsmittel in der Kupfer–und Bronzezeit Mitteleuropas. Fundberichte aus Baden–Württemberg 20: 229–327.
—. 2002: M. Lenerz De–Wilde, Bronzezeitliche Zahlungsmittel. Mitteilungen der Anthropologischen Gesellschaft in Wien. Band 132: 1–23.
Lenneis et al. 1995: E. Lenneis, Ch. Neugebauer–Maresch, E. Ruttkay, Jungsteinzeit im Osten Österreichs. Wien: Verlag Niederösterreichisches Pressehaus.
Lenneis 2000: E. Lenneis, Hausformen der mitteleuropäischen Linearbandkeramik und des balkanischen Frühneolithikums im Vergleich. In: St. Hiller, V. Nikolov (eds.), Karanovo III: Beiträge zum Neolithikum in Südosteuropa: 383–388. Wien: Phoibos.
Levi–Strauss 1991: C. Lévi–Strauss, Maison. In: P. Bonte, M. Izard (eds.), Dictionnaire de l'ethnologie et de l'anthropologie: 434–436. Paris.

Lewis–Williams & Pearce 2009: D. Lewis–Williams, D. Pearce, Inside the Neolithic Mind. Consciousness, Cosmos and the Realm of the Gods. London: Thames and Hudson.

Lichardus 1981: J. Lichardus, Zur Bedeutung der Feuersteingewinnung in der jüngeren Steinzeit Mitteleuropas. In: 5000 Jahre Feuersteinbergbau: 265–270. Bochum: Veröffentlichung der Bergbaumuseum.

Lichardus–Itten 1981: M. Lichardus–Itten, Silexknollen als Beigabe der frühkupferzeitlichen Tiszapolgar–Kultur. In: 5000 Jahre Feuersteinbergbau: 279–283. Bochum: Veröffentlichung der Bergbaumuseum.

Lichter 2001: C. Lichter, Untersuchungen zu den Bestattungssitten des südosteuropäischen Neolithikums und Chalkolithikums. Mainz am Rhein: Zabern Verlag.

Lüning 1997: J. Lüning, Haus, Hof und Siedlung im älteren Neolithikum am Beispiel der Aldenhovener Platte in der Niederrheinischen Bucht. In: H. Beck/H. Steuer (Hrsg.), Haus und Hof in ur–und frühgeschichtlicher Zeit. Abh. Akad. Wiss. Göttingen, Phil.–Hist. Kl. 3. Folge 128: 70–75.

—. 2005: J. Lüning, Zwischen Alltagswissen und Wissenschaft im Neolithikum. In: T. L. Kienlin (hrsg.), Die Dinge als Zeichen: Kulturelles Wissen und materielle Kultur: 53–80. Bonn: Habelt Verlag.

Lyotard 1979: J.–F. Lyotard, La condition postmoderne: rapport sur le savoir. Paris.

Malafouris 2004: L. Malafouris, The Cognitive Basis of Material Engagement: Where Brain, Body and Culture Conflate. In: E. DeMarrais, Ch. Gosden, C. Renfrew (eds.), Rethinking materiality. The engangement of mind with the material world: 53–62. Cambridge: Cambridge University Press.

—. 2010: L. Malafouris, Grasping the concept of number: How did the sapient mind move beyond approximation? In: I. Morley, C. Renfrew (eds.), The Archaeology of Measurement. Comprehending Heaven, Earth and Time in Ancient Societies: 35–42. Cambridge: Cambridge University Press.

Manolakakis 2005: L. Manolakakis, Les industries lithiques énéolithiques de Bulgarie. Rahden/Westf: Verlag Marie L. Leidorf.

Marshack 1972: A. Marshack, Roots of Civilization: Cognitive Beginnings of Man's First Art Symbol and Notation. Littlehampton: McGraw–Hill.

—. 1972a: A. Marshack, Upper Paleolithic Notation and Symbol. Sequential microscopic analyses of Magdalenian. Science, New Series, Vol. 178: 817–828.

Marshall 1981: A. Marshall, Environmental adaptation and structural design in axially–pitched longhouses from Neolithic Europe. World Archaeology Vol. 13, No. 1: 101–121.
Mayer 1977: E. F. Mayer, Die Äxte und Beile in Österreich. PBF IX, 9 Band. München.
McLuhan 1964: M. McLuhan, Understanding Media. The Extensions of Man. New York: New American Library.
—. 2004: M. McLuhan, Zrozumieć media. Przedłużenie człowieka. Warszawa: Wyd. Naukowo–Techniczne.
Mellaart 1967: J. Mellaart, Çatal Hűyűk: A Neolithic Town in Anatolia. London–New York: Thames, Hudson.
Merleau–Ponty 1945: M. Merleau–Ponty, Phénoménologie de la perception. Paris: Gallimard.
—. 2001: M. Merleau–Ponty, Fenomenologia percepcji. Warszawa: Aletheia.
Merleau–Ponty 2002: M. Merleau–Ponty, Phenomenology of Perception. Routledge.
Merlini 2006: M. Merlini, Semiotic approach to the features of the "Danube Script". Documenta Praehistorica XXXII: 233–251.
—. 2009: M. Merlini, An Inquiry into the Danube Script. Sibiu–Alba Iulia: Ed. Altip.
Michałowski 1993: P. Michalowski, Tokenism. American Anthropologist, New Series, Vol. 95, No. 4: 996–999.
Midgley 2005: M. Midgley, Monumental Cemeteries of Prehistoric Europe. Tempus.
Migal 2002: W. Migal, Zamysł technologiczny wióra krzemiennego z Winiar, gm. Dwikozy. In: B. Matraszek, S. Sałaciński (eds.), Krzemień świeciechowski w pradziejach. Studia nad gospodarką surowcami krzemiennymi w pradziejach 4: 255–266. Warszawa.
Milisauskas & Kruk 1989: S. Milisauskas, J. Kruk, Economy, migration, settlement organisation, and warfare during the late Neolithic in Southeastern Poland. Germania t. 67, nr 1: 77–96.
Milisauskas 2002: S. Milisauskas, Early Neolithic, The First Farmers in Europe, 7000–5500/5000 BC. In: (red.) S. Milisauskas, European Prehistory. A Survey: 153–206. New York–Moscow: Springer.
Miller 1992: K. Miller, What a Number is: Mathematical Foundations and Developing a Number Concept. In: G . E. Stelmach, P. A. Vroon (eds.), The Nature and Origins of Mathematical Skills: 3–38. Amsterdam–Tokyo: North–Holland 1992.
Mithen 1996: S. Mithen, The Prehistory of the Mind. A search for the origins of art, religion and science. London: Phoenix.

Modermann 1986: P.J.R. Modermann, On the typology of the houseplans and their europaean setting. In: I. Pavlu, J. Rulf, M. Zapotocka (eds.), Theses on the Neolithic site of Bylany. Pamatky Archeologicke 77: 383–394.

—. 1988: P.J.R. Modermann, The Linear Pottery Culture: Diversity in Uniformity. Berichten van de Rijksdienst voor Outheidkundig Bodemonderzoek 38: 63–139.

Morgan 1877: L.H. Morgan, Ancient Society, or Researches in the Lines of Human Progress from Savagery, through Barbarism, to Civilization. New York.

Müeller 2001: J. Müller, Soziochronologische Studien zum Jung–und Spätneolithikum im Mittelelbe–Saale–Gebiet (4100–2700 v. Chr). Rahden/Westf.: Verlag Marie L. Leidorf.

—. 2005: J. Müller, Zur Rolle von Alter und Geschlecht in neolithischen Gesellschaften Europas. In: Müller (hrsg.), Alter und Geschlecht in ur– und frühgeschichtlichen Gesellschaften: 19–25. Universitätsforschungen zur Prähistorischen Archäologie, Band 126. Bonn: Verlag Dr. Rudolf Habelt.

Müeller–Beck 2001: H. Müller–Beck, Eiszeitkunst im Süddeutsch.–Schweizerischen Jura. In: Müller–Beck, H.–J. / Conard, N. J. (Hrsg.), Eiszeitkunst im Süddeutsch–Schweizerischen Jura. Stuttgart: Theiss Verlag.

Müller et al. 1996: J. Müller, J., Herrera A., N. Knossalla, Spondylus und Dechsel–zwei gegensätzliche Hinweise auf Prestige in der mitteleuropäischen Linearbandkeramik?. In: J. Müller, R. Bernbeck (hrsg.), Prestige–Prestigegüter–Sozialstrukturen. Beispiele aus dem europäischen und vorderasiatischen Neolithikum: 81–96. Archäologische Berichte 6. Bonn: Habelt Verlag.

Neugebauer 2002: J–W. Neugebauer, Die Metalldepots der Unterwölblinger Kulturgruppe Ragelsdorf 2 und Unterradlberg 1 und 2. Überlegungen zum prämonetären Charakter der niedergelegten Wertgegenstände. Mitteilungen der Anthropologischen Gesellschaft in Wien 132: 25–40.

Nielsen 1977: O. Nielsen, Die Flintbeile der Frühen Trichterbecherkultur in Daenemark. Acta Archaeologica 48.

Nieszery 1995: N. Nieszery, Linearbandkeramische Gräberfelder in Bayern. Internationale Archäologie Band 16. Verlag Marie L. Leidorf.

Nikolov 1991: V. Nikolov, Längenbaumaß im Frühneolithikum. Archäologisches Korrespondenzblatt 21: 45–48.

—. 1998: V. Nikolov, Das megalithische Yard–Baumaß in den Hausbauten der bulgarischen Tellsiedlungen. In: D. Ahrens/R.h Rottländer (Hrsg.),

Internationaler interdisziplinärer Kongreß für Historische Metrologie: 115–124. St. Katharinen: Scripta Mercaturae.
Nissen 1988: H. J. Nissen, The Early History of the Ancient Near East 9000–2000 BC. Chicago–London: Chicago University Press.
Nissen et al. 1993: H. J. Nissen, P. Damerow, R. K. Englund, Archaic Bookkeeping: Early Writing and Techniques of Economic Administration in the Ancient Near East. Chicago: Chicago University Press.
Noble 1999: D.F. Noble, The Religion of Technology. The Divinity of Man and the Spirit of Invention. Penguin Books.
Oates 1993: J. Oates, Trade and Power in the fifth and fourth millenia BC: new evidence from northern Mesopotamia. World Archaeology, Volume 24, No. 3: 403–422.
Olausson 1983: D. Olausson, Lithic Technological Analysis of the Thin–Butted Flint Axe. Acta Archaeologica 53.
—. 1997: D. Olausson, Craft specialization as an agent of social power in the south Scandinavian Neolithic. In: R. Schild, Z. Sulgostowska (eds), Man and Flint. Proceedings of the 7th International Flint Symposium: 269–278. Warszawa.
Ong 2009: W. J. Ong, Osoba, świadomość, komunikacja. Warszawa: Wydawnictwo Uniwersytetu Warszawskiego.
Ottaway & Strahm 1975: B. Ottaway, Chr. Strahm, Swiss Neolithic copper beads: currency, ornament or prestige items?. World Archaeology 6: 307–321.
Pare 1999: Chr. Pare, Weights and Weighting in Bronze Age Central Europe. In: Eliten der Bronzezeit. Ergebnisse zweier Kolloquien in Mainz und Athen. Mainz: Verlag der Römisch–Germanischen Zentralmuseums.
Pavlu 2000: I. Pavlu, Life on a Neolithic site: Bylany–situational analysis of artefacts. Prague: Institute of Archaeology Press.
—. 1972: J. Pavuk, Neolithiches Gräberfeld in Nitra. Slovenska Archeologia XX: 5–105.
Pechtl 2009: J. Pechtl, A monumental prestige patchwork. In: D. Hofmann & P. Bickle (eds.), Creating communities: new advances in Central European Neolithic Research: 186–201. Oxford: Oxbow.
Pellegrin 2006: J. Pellegrin, Long blade technology in the Old World: an experimental approach and some archaeological results. In: (red.) J. Apel, K. Knutsson, Skilled Production and Social Reproduction: 37–68. Upsalla: Societas Archeologica Upsaliensis.

Peroni 1998: R. Peroni, Bronzezeitliche Gewichtssysteme im Metallhandel zwischen Mittelmeer und Ostsee. In: B. Hänsel (red.), Mensch und Umwelt in der Bronzezeit Europas. Kiel: Oetker–Voges Verlag.

Petrasch 1990: J. Petrasch, Mittelneolithische Kreisgrabenanlagen in Mitteleuropa. Bericht der Römisch–Germanischen Komission, t. 71: 407–564.

Petrequin et al. 1998: Petrequin, P., A.–M. Petrequin, F. Jeudy, C. Jeunesse, J.–L. Monnier, J. Pelegrin, I. Praud, From the Raw Material to the Neolithic Stone Axe. Production Process and Social Context. In: M. Edmonds, C. Richards (eds.), Understanding the Neolithic of North Western Europe: 277–311. Glasgow: Cruithne Press.

Petruso 1992: K. M. Petruso, Ayia Irini: The Balance Weights. An Analysis of Wheight Measurement in Prehistoric Crete and the Cycladic Islands. Mainz am Rhein: Philipp von Zabern Verlag.

Pigot 1983: S. Pigott, The Earliest Wheeled Transport: From the Atlantic Coast to the Caspian See. Ithaca.

Plato, The Republic.

Podborsky et al. 2002: V. Podborsky a kolektiv, Dve pohrebiste neolitickeho lidu s linearni keramikou ve Vedrovicich na Morave. [Zwei Gräbelfelder des neolithischen Volkes mit Linearbandkeramik in Vedrovice in Mähren]. Brno: Institute of Archaeology.

Polmann 2003: T. Polmann, Some Principles Involved in the Acquisition of Number Words. Language Acquisition 11(1): 1–31.

Putnam 1973: H. Putnam, Meaning and Reference. Journal of Philosophy 70: 699–711.

Raczky et al. 1996: P. Raczky et. all., Two unique assemblages from the Late Neolithic tell settlement at Polgar–Csoszhalom. In: T. Kovacs (hrsg.), Studien zur Metallindustrie im Karpatenbecken und den benachbarten Regionen. Festschrift für Amalia Mozsolics zum 85. Geburtstag: 17–31. Budapest: Magyar Nemzeti Muzeum.

Raczky et al. 1994: Raczky et all., Polgar–Csoszhalom. A late Neolithic Settlement in the Upper Tisza region and its cultural connections (Preliminary report). Jósa András Múzeum Évkönyve 36: 231–240.

Raetzel–Fabian 2000: D. Raetzel–Fabian, Calden. Erdwerk und Bestattungsplätze des Jungneolithikums. Architektur–Ritual–Chronologie. Universitätsforschungen zur Prähistorischen Archäologie, Band 70. Bonn: Rudolf Habelt Verlag.

Rahmstorf 2010: L. Rahmstorf, The concept of weighing during the Bronze Age in the Aegean, the Near East and Europe. In: I. Morley, C. Renfrew (eds.), The Archaeology of Measurement. Comprehending

Heaven, Earth and Time in Ancient Societies: 87–105. Cambridge 2010.
Rajchl 2002: R. Rajchl, Archeoastronomicka analyza orientace skeletu na pohrebisti v „Siroke u lesa". In: V. Podborsky et all. 2002, Dve pohrebiste neolitickeho lidu s linearni keramikou ve Vedrovicich na Morave. [Zwei Graebelfelder des neolithischen Volkes mit Linearbandkeramik in Vedrovice in Mähren: 275–292. Brno: Institute of Archaeology.
Rappenglück 2012: M. Rappenglück, The Decorated Plate of the Geißenklösterle, Germany, http://www.astronomicalheritage.org/index.php?option=com_content&view=article&id=28&Itemid=33
Rasch 1987: W. Rasch, Gab es im Neolithikum ein einheitliches Baumaß? Archäologisches Korrespondenzblatt 17: 341–346.
—. 2000: W. Rasch, Massversuche an Grundrissen neolithischer Hausbauten der bulgarischen Tellsiedlungen. Beiträge zur Ur–und Frühgeschichte Mitteleuropas 22: 33–46.
Rech 1979: M. Rech, Studien zu Depotfunden der Trichterbecher–und Einzelgrabkultur des Nordens. Offa–Bücher 39. Neumünster: Wachholz.
Renfrew & Bahn 1990: C. Renfrew, P. Bahn, Archaeology: Theories, Methods and Practice. Thames, Hudson.
Renfrew & Bahn 2002: C. Renfrew, P. Bahn, Archeologia. Teorie, metody, praktyka. Warszawa: Prószyński, S–ka.
Renfrew 1976: C. Renfrew, Before Civilization. The Radiocarbon revolution and Prehistoric Europe. Pimlico 1976.
—. 1976a: C. Renfrew, Megaliths, territories and populations. In: S. J. De Laet (ed.), Acculturation and Continuity in Atlantic Europe: 198–220. Brugge: De Tempel.
—. 1990: C. Renfrew, Archaeology and Language. The Puzzle of Indo–Europaean Origins. Cambridge: Cambridge University Press.
—. 2001: C. Renfrew, Język i archeologia, Warszawa–Poznań: Wydawnictwo Naukowe PWN.
—. 2004: C. Renfrew, Towards a Theory of Material Engagement. In: E. DeMarrais, Ch. Gosden, C. Renfrew (eds.), Rethinking materiality. The engangement of mind with the material world: 23–32. Cambridge 2004.
—. 2007: C. Renfrew, Prehistory. The Making of the Human Mind. London–New York: Phoenix.
Romer 2007: J. Romer, The Great Pyramid: ancient Egypt revisited. Cambridge University Press.

Rorty 1989: R. Rorty, Contingency, Irony, and Solidarity. Cambridge: Cambridge University Press.
Rosenberg 2007: D. Rosenberg, Making Time. Cabinet Magazine, Issue 28 Winter 2007/08, http://www.cabinetmagazine.org/issues/28/rosenberg.php
Rottländer 1999: R. Rottländer, Die Maßeinheit der Bandkeramik vor dem Hintergrund der antiken Metrologie. Archäologisches Korrespondenzblatt 29: 189–202.
Roux 1985: G. Roux, La Mesopotamie. Paris: Seuil.
—. 2003: G. Roux, Mezopotamia. Warszawa: Wydawnictwo Akademickie DIALOG.
Ruggles 2012: C. Ruggles, Later Prehistoric Europe, Heritage Sites of Astronomy and Archaeoastronomy in the context of the Unesco World Heritage Convention, http://www2.astronomicalheritage.org/index.php/thematic–study
Rück 2009: O. Rück, New aspects and model for Bandkeramik settlement research. In: D. Hofmann, P. Bickle (eds.) Creating communities: new advances in Central European Neolithic Research: 159–185. Oxford: Oxbow.
Rzepecki 2011: S. Rzepecki, The Roots of Megalithism in the TRB Culture. Łódź: Institute of Archaeology.
Sangmeister & Strahm 1974: E. Sangmeister, Chr. Strahm, Die Funde aus Kupfer in Seeberg, Burgäschisee–Süd. Acta Bernensia II.
Sawday 2007: J. Sawday, Engines of the Imagination. Renaissance culture and the rise of the machine. London–New York: Routledge.
Scarre 1992: Ch. Scarre, The Early Neolitic of Western France and Megalithic Origins in Atlantic Europe. Oxford Journal of Archaeology 11(2): 121–154.
—. 2007: Ch. Scarre, Changing Places: monuments and the Neolithic transition in western France. Proceedings of the British Academy 144: 243–262.
Schiffer 1999: M.B. Schiffer, The Material Life of Human Beings. Artifact, behavior and communication. London–New York: Taylor, Francis.
Schmandt–Besserat 1982: D. Schmandt–Besserat, The Emergence of Recording. American Anthropologists 84: 871–878.
—. 1992: D. Schmandt–Besserat, Before Writting. From Counting to Cuneiform. Austin: University of Texas Press.
—. 2007: D. Schmandt–Besserat, Jak powstało pismo. Warszawa: AGADE.

—. 2009: D. Schmandt–Besserat, Tokens and Writing: the Cognitive Development. SCRIPTA, Volume 1: 145–154.

—. 2010: D. Schmandt–Besserat, The token system of the ancient Near East: Its role in counting, writing, the economy and cognition. In: I. Morley, C. Renfrew (eds.), The Archaeology of Measurement. Comprehending Heaven, Earth and Time in Ancient Societies: 27–34. Cambridge: Cambridge University Press.

Schmidt 2006: K. Schmidt, Sie bauten die ersten Tempel. Das rätselhafte Heiligtum der Steinzeitjäger. München: C.H. Beck Verlag.

—. 2004: A. Schmitz, Typologische, chronologische und paleometallurgische Untersuchungen zu den fruehkupferzeitlichen Kupferflachbeilen und Kupfermeisseln in Alteuropa. Unpublizierte Doktorarbeit. Saarbruecken.

Schwartz 1994: G. M. Schwartz, Before Ebla: Models of Pre–State Political Organization in Syria and Northern Mesopotamia. Monographs in World Archaeology No. 18: 153–174.

Seferiades 2000: M.L. Seferiades, Spondylus Gaederopus: Some Observations on the Earliest Europaean Long Distance Exchange System. Spondylus Gaederopus: Some observations on the earliest European long distance exchange system. In: Karanovo, Band III: Beiträge zum Neolithikum in Südosteuropa (Hrsg. S. Hiller, V. Nikolov): 423–437. Wien: Phoibos Verlag.

Shennan 1993: S. Shennan, Settlement and Social Change in Central Europe 3500–1500 BC. Journal of World Prehistory, Vol. 7, No. 2: 121–161.

Sherratt 1982: A. Sherratt, Mobile resources: settlement and exchange in early agricultural Europe. In: C. Renfrew, S. Shennan (eds.), Ranking, Resource and Exchange: 27–32. Cambridge: Cambridge University Press.

—. 1986: A. Sherratt, Cups that cheered. In: W. Waldren, R. Kennard, (eds.), Bell Beakers of the Western Mediterranean (British Archaeological Reports SS 331): 81–114.

—. 1990: A. Sherratt, The genesis of megaliths: monumentality, ethnicity and social complexity in Neolithic north–west Europe. World Archeology 22(2): 147–167.

—. 1994: A. Sherratt, The emergence of elites: earlier Bronze Age Europe 2500–1300 BC. In: B. Cunliffe (ed.), The Oxford Illustrated Prehistory of Europe: 244–276. Oxford: Oxford University Press.

—. 1997: A. Sherratt, Economy and Society in Prehistoric Europe. Edinburg: Edinburg University Press.

Šiška 1964: S. Šiška, Pohrebisko tiszapolgarskiej kultury w Tibave, Slovenska Archeologia 12: 293–356.
Sloterdijk 2005: P. Sloterdijk, Im Weltinnenraum des Kapitals: für eine philosophische Theorie der Globalisierung. Suhrkamp.
Soin 2001: M. Soin, Gramatyka i metafizyka. Problem Wittgensteina. Wrocław: Wydawnictwo Uniwersytetu Wrocławskiego.
Sommerfeld 1994: Chr. Sommerfeld, Gerätegeld Sichel. Studien zur monetären Struktur bronzezeitlicher Horte im nördlichen Mitteleuropa. Berlin–New York: Walter de Gruyter.
Staaf 1996: B. M. Staaf, An Essay on Copper Flat Axes. Acta archaeologica Lundensia N. 21.
Starovič 2006: A. Starovič, If the vinča script once really existed who could written or read it? Documenta Praehistorica XXXII: 253–260.
Startin 1978: W. Startin, Linear Pottery Culture Houses: Reconstruction and Manpower. Proceedings of the Prehistoric Society 44: 143–159.
Stein 1994: G. Stein, Economy, Ritual, and Power in Ubaid Mesopotamia. In: Chiefdoms and Early States in the Near East: The Organizational Dynamics of Complexity. Monographs in World Archaeology, No. 18: 35–46.
Stout 2002: D. Stout, Skill and Cognition in Stone Tool Production. An Ethnographic Case Study from Irian Jaya. Current Anthropology Vol. 45: 693–722.
Straffin 1993: P. D. Straffin, Game Theory and Strategy. Washington: MAA.
Strahm 1992: Chr. Strahm, Die Dynamik der schnurkeramischen Entwicklung in der Schweiz und in Südwestdeutschland. Schnurkeramik Symposium 1990. Praehistorica XIX: 187–198.
—. 1994: Chr. Strahm, Anfänge der Metallurgie in Mitteleuropa. Helvetia Archaeologica 25: 2–39.
—. 2004: Chr. Strahm, Die Glockenbecher–Phänomen aus der Sicht Komplementär–Keramik. In: J. Czebreszuk (Hrsg.), Similar but Different. Bell Beakers in Europe: 101–126. Poznań: Instytut Archeologii.
Szmyt 1996: M. Szmyt, Społeczności Kultury Amfor Kulistych na Kujawach. Poznań: Instytut Archeologii.
Tacon 1991: P.S.C. Tacon, Aspects of stone use and toll development in Western Arnhem Land. Antiquity 65: 192–207.
Tarnas 1993: R. Tarnas, The Passion of the Western Mind. Random House Publishing.
Thom 1955: A. Thom, Journal of the Royal Statistical Society. Series A (General) Vol. 118, No. 3: 275–295.

—. 1967: A. Thom 1967, Megalithic Sites in Britain. Oxford: Oxford University Press.
Thomas 1987: J. Thomas, Relations of Productions and Social Change in the Neolithic of North–Wets Europe. Man New Series, Vol. 22, No. 3: 405–430.
—. 1991: J. Thomas, Rethinking the Neolithic. Cambridge: Cambridge University Press.
Tilley 1994: Ch. Tilley, A phenomenology of landscape: places, paths, and monuments. Berg Publishers.
—. 1996: C. Tilley, An Ethnology of the Neolithic. Cambridge: Cambridge University Press.
—. 1999: Ch. Tilley, Metaphor and Material Culture. Blackwell.
Todorova 1978: H. Todorova, The Eneolithic in Bulgaria. Oxford: British Archaeological Reports.
Tomasello 1999: M. Tomasello, The Cultural Origins of Human Cognition. Harvard University Press.
—. 2002: M. Tomasello, Kulturowe źródła ludzkiego poznania. Warszawa: Państwowy Instytut Wydawniczy.
Treherne 1995: P. Treherne, The Warriors Beauty: the masculine body and self–identity in Bronze–Age Europe. Journal of European Archaeology, vol. 3:1: 105–144.
Trick 1992: L. M. Trick, A Theory of Enumeration that grows out of a general Theory of Vision: Subitizing, Counting, and Finsts. In: G.E. Stelmach, P. A. Vroon (eds.), The Nature and Origins of Mathematical Skills: 257–299. Amsterdam–Tokyo: North–Holland.
Tringham 1971: R. E. Tringham, Hunters, Fishers and Farmers of Eastern Europe 6000–3000 BC. London: Hutchinson.
Trnka 1990: G. Trnka, Zum Forschungsstand der mittelneolithischen Kreisgrabenanlagen in Oesterreich. Jahresschrift für die Mitteldeutsche Vorgeschichte 73: 7–14.
Tsonev 2004: T. Tsonev, Long blades/superblades in Anatolian and east Balkan Neolithic contexts. British Archaeological Reports (IS) 1303: 17–24.
Tunia 2003: K. Tunia, Słonowickie megaksylony. Archeologia Żywa, nr 1.
Turek 2001: J. Turek, Stone axes as tools, valuables and symbols (3300–1900 BC). In: D. Georghiu (ed.), Material, Virtual and Temporal Compositions: On the Relationship between Objects, British Archaeological Reports 953 (IS): 53–62.
Vait 1999: U. Vait, Überlegungen zur Funktion und Bedeutung der Megalithgraeber im noerdlichen und westlichen Europa. Beiträge zur

Ur–und Frühgeschichte Mitteleuropas 21, Studien zur Megalithik: 395–419. Weissbach: Beier, Beran.
Vial 1940: L. Vial, Stone Axes of Mount Hagen, New Gwinea. Oceania, Vol. XI, no. 2: 158–163.
Videiko 2009: M. Videiko, Signs and Sign Systems of the Trypillia Culture (5400–2700 BC). In: Signs of Civilzation. Neolithic Symbol System of Southeast Europe. Proceedings from the International Symposium Signs of Civilization–Novi Sad, Serbia: 179–186.
Voss 1982: J.A. Voss, A Study of Western TRB Social Organization. Ber. Rijksd. Oudh. Bodemond, Jrg. 32.
Watkins 2004: T. Watkins, Building houses, framing concepts, constructing worlds. Paléorient 30(1): 5–24.
—. 2006: T. Watkins, Architecture and the symbolic construction of new worlds. In: E.B. Banning, M. Chazan (eds.), Domesticating space: construction, community and cosmology in the late prehistoric Near East: 15–24. Berlin: Ex Oriente.
Weiner 2003: J. Weiner, Profane Geräte oder Prunkstücke? Überlegungen zur Zweckbestimmung übergrosser Dechselklingen. In: J. Eckert, U. Eisenhauer, A. Zimmermann (Hrsg.), Studia honoraria. Internationale Archäologie 20: 423–440. Rahden.
Wengrow 1998: D. Wengrow, The changing face of clay: continuity and change in the transition from village to urban life in the Near East. Antiquity 72: 783–795.
—. 2010: D. Wengrow, What Makes Civilization. The Ancient Near East and the Future of the West. Oxford University Press.
West 1999: M.L. West, The East Face of Helicon. West Asiatic Elements in Greek Poetry and Myth. Oxford University Press.
Whittle 1995: A. Whittle, Gifts from the earth: symbolic dimensions of the use and production of Neolithic flint and stone axes. Archeologia Polona, vol. 33: 247–259.
—. 1996: A. Whittle, Houses in context: Buildings as process. In: T. Darvill, J. Thomas (eds.), Neolithic Houses in Northwest Europe and Beyond: 13–26. Oxford: Oxbow Books.
Whorf 1956: In J. B. Carroll (ed.), Language, thought, and reality: Selected writings of Benjamin Lee Whorf. Cambridge: MIT Press.
—. 2002: B. Lee Whorf, Język, myśl, rzeczywistość. Warszawa: Wydawnictwo KR.
Wierciński 1994: A. Wierciński, Magia i religia. Szkice z antropologi religii. Kraków: Nomos.
Wilson 1988: P. Wilson, The Domestication of the Human Species. Yale University Press.

Winiger 1985: J. Winiger, Das Neolithikum der Schweiz: eine Vorlesungsreihe zum Forschungsstand. Basel: Sem. für Ur– und Frühgeschichte.

Winn 2009: S.M. Winn, The Danube (Old Europaean) Script. Ritual use of signs in the Balkan–Danube Region c. 5200–3500 BC. In: Signs of Civilzation. Neolithic Symbol System of Southeast Europe. Proceedings from the International Symposium Signs of Civilization–Novi Sad, Serbia 2004: 49–62.

Wiślański 1969: T. Wiślański, Podstawy gospodarcze plemion neolitycznych w Polsce północno–zachodniej. Wrocław–Warszawa–Kraków: Zakład Narodowy im. Ossolińskich.

Witter 1941: W. Witter, Neue Ergebnisse der Metallforschung in Deutschland und ihre Beziehung zu den Schnurkaramikern. Mannus 33: 70–80.

Wittgenstein 1974: L. Wittgenstein 1974, Philosophical Grammar. Berkeley–Los Angeles: University of California Press.

Zakościelna 2010: A. Zakościelna, Studium obrządku pogrzebowego kultury lubelsko–wołyńskiej. Lublin: Wydawnictwo Naukowe UMCS.

Zerries & Schuster 1974: O. Zerries, M. Schuster, Mahekodotedi. Monographie eines Dorfes der Waika–Indianer (Yanomama) am Oberen Orinoko (Venezuela). München: Klaus Renner Verlag.

Zvelebil 2001: M. Zvelebil, The Agricultural Transition and the origins of the Neolithic Society in Europe. Documenta Praehistorica XXVIII: 1–26.

—. 2005: M. Zvelebil, Homo habitus: agency, structure and the transformation of tradition in the constitution in the TRB foraging-farming communities in the North Europaean plain (ca. 4500–2000 BC). Documenta Praehistorica XXXII: 87–101.

Žižek 1992: S. Žižek, Looking Awry. An Introduction to Jacues Lacan through Popular Culture. MIT Press.

—. 2003: S. Žižek, Patrząc z ukosa. Do Lacana przez kulturę popularną. Warszawa: Wydawnictwo KR.

INDEX

abstract thinking, 43, 48
abstract value, 131
accumulation, 131, 159, 177, 184, 189
administration, 21, 36, 42, 46, 49, 154
adorant, 19
agglomeration, 64, 176
agglomerative, 64, 66
agrios, 75
alcohol, 93, 156
Alps, 62, 70, 124, 134, 136, 159, 190
alter ego, 159, 172
ancestors, 15, 24, 57, 59, 82, 87, 93, 100, 114, 126, 128, 145, 165, 180, 196
animals, 14, 25, 35, 40, 43, 50, 52, 56, 74, 110, 157, 181, 184, 188
anthropogenic, xii, 26, 65, 90, 94, 97, 183
anthropologic, 110, 152, 183
anthropomorphic, 19, 32, 41, 56, 147
anthropomorphic figure, 19, 41, 56
architecture, xiii, 39, 63, 66, 71, 74, 76, 85, 96, 100, 119, 125, 156, 168, 172, 174, 177, 186, 195
aristocracy, 190
astronomy, 14, 18, 68, 86, 91, 103, 128, 197
axis mundi, 175, 179, 183
Balkan scripture, 54, 174
Balkans, 53, 62, 64, 72, 76, 80, 88, 97, 100, 103, 122, 159, 161, 174
bars
    metal bars, 140, 143, 145, 159, 161, 164, 189, 197
    axe bars, 159, 162, 191, 197

beads, xiii, 104, 134, 154, 162, 164, 187, 197
bell beaker, 149, 154, 157, 162, 188
blade fragments, 124, 126, 172
brain research, 15, 31
Broadbent, 138, 150
bronze, xiv, 81, 86, 93, 100, 131, 146, 154, 158, 162, 165, 170, 174, 181, 186
Bronze Age, xiv, 81, 86, 93, 100, 131, 146, 154, 158, 162, 165, 170, 174, 181, 186
calculi, 20
calendar, 18, 55, 93, 103
Carpathian Basin, 53, 77, 84, 101, 103, 123, 127, 146
Çatal Hűyűk, 64, 98, 175, 177
caves, x, 17, 20, 26, 81, 97, 147, 180, 195
Cemeteries, 80, 82, 99, 101, 105, 118, 122, 155, 158, 177, 185, 195
ceramics,
    amphorae, 116, 125, 149, 154, 193
    beakers, 41, 47, 63, 93, 117, 125, 127, 149, 154, 162, 188, 190, 193, 197
    Begleitkeramik, 149
    cardium pottery, 53
    ceramic vessels, xiv, 177, 188
    clay, xii, 22, 35, 40, 43, 45, 52, 56, 62, 67, 89, 152
    clay cones, 52
    clay stamps, 40, 52, 62, 178
    containers, 24, 36, 48, 58, 170
    corded ware, 149, 158, 188, 197
    envelopes, 41, 43, 58
    painted pottery, 53, 64

pottery, 36, 39, 41, 53, 55, 62, 64, 72, 83, 106, 109, 131, 142, 147, 152, 177, 181, 189, 192
   vessels, xiv, 25, 36, 39, 46, 55, 62, 101, 104, 106, 123, 147, 152, 170, 177, 181, 188, 197
   zoomorphic figures, 53
ceremonial, 56, 68, 77, 83, 105, 113, 130, 159, 165
ceremonies, 18, 39, 57, 82, 122, 127
Cerny, 79, 85
Chalkolithic, 67, 77, 121
chiefdom, 189, 193
child, v, 1, 8, 15, 19, 25, 42, 79, 101, 103, 118, 149, 155
Chomsky, 9
   language recursiveness, 10
circles, xii, 46, 86, 92, 94
circular motion, 182
cognition,
   cognitive, 1, 3, 21, 23, 29, 51, 70, 91, 96, 98, 154, 165, 167, 170, 180, 188, 195, 198
   cognitive abilities, 4, 7, 23, 188
   cognitive fluidity, 24, 180
Colmar, vi, ix, xiii, 141, 143, 146, 187
colonization of space, 70
communal, 64, 83, 98, 127, 156, 158, 177, 180, 194
communication, xiii, 34, 38, 41, 47, 51, 53, 57, 68, 84, 95, 100, 114, 121, 128, 130, 134, 139, 145, 154, 159, 166, 168, 171, 178, 181, 187, 194
   avoiding misunderstanding, 7
   communicative actions (theory), 7
   communication medium, xiii, 97, 100
   communication system, 38, 57, 96, 178
   discourse, 7, 76, 79, 94, 103, 115, 125, 145, 156, 161, 163, 171, 180, 193
   gestures and rituals, 8
   human communication, 6, 8, 34, 47, 166, 171
   media, 8, 30, 33, 51, 63, 97, 127, 129, 154, 168, 178, 180, 185, 197
   mimetical communication, 8
   reaching agreements, 8
   social cooperation, 7
   steering media, 8
   verbal communication, 8
   verbalization of sacrum, 51
community, xii, 59, 63, 65, 68, 74, 79, 82, 84, 98, 101, 114, 125, 127, 130, 173, 175, 178, 181, 184, 186, 189, 192, 197
community cohesion, 65
constructed environment, 65
consumption, 114, 157, 189
control, 18, 26, 34, 36, 38, 40, 44, 63, 75, 81, 97, 116, 121, 131, 157, 173, 175, 177, 180, 189
cooperation, 7, 90, 115, 168, 180
copper, xiii, 56, 103, 119, 121, 125, 129, 134, 138, 154, 158, 169, 181, 185, 189, 191, 197
Cortaillod, xiii, 134, 139, 145, 164
cosmic order, xii, 68
cosmologic beliefs, 26, 101
cosmology, xii, 18, 26, 68, 70, 74, 77, 81, 85, 96, 100, 127, 178, 182, 188, 197
cosmos, 68, 81, 97, 126, 179, 182
counting, 1, 3, 7, 22, 31, 35, 40, 42, 47, 53, 56, 58, 63, 87, 98, 100, 118, 123, 128, 131, 137, 142, 147, 159, 169, 173, 176, 184, 189, 192
   concrete counting, xi, 29, 42, 50, 98, 108, 140, 146, 165, 186
   counting system, 11, 42, 48, 170, 173
   decimal, 48, 87, 94, 174
   one to one, 16, 20, 42, 69, 100, 147

quaternary system, 170
quinary system, 170
sexagesimal, 49, 51
vigesimal, 49, 174
crystal palace, 183
culture, 4, 8, 16, 20, 26, 30, 32, 34, 37, 39, 50, 59, 62, 69, 75, 78, 92, 95, 101, 104, 106, 109, 116, 120, 124, 131, 134, 141, 149, 154, 167, 174, 177, 180, 183, 186, 188
Danube, 54, 62, 84
deposit, 36, 102, 105, 109, 112, 121, 123, 125, 130, 137, 140, 147, 158, 160, 186, 189, 194
Descartes, 6, 29
dialectic, 124, 131, 149, 187
digits, 33, 42, 45, 60, 172
discretization, 33
Dolni Věstonice, 18
domestication, xii, 56, 75, 99
  domestic life, 42
  domesticated communities, 65
  domus, 74, 78, 156, 180, 183, 196
egalitarian, 68, 158, 184, 188
enchainment, 106, 131, 159, 172, 177, 187, 197
Eneolithic, xii, 27, 55, 77, 82, 84, 89, 93, 96, 98, 103, 108, 114, 119, 123, 130, 132, 137, 140, 145, 148, 152, 169, 176, 180, 182, 188, 190, 194
ethnology, 10, 12, 15, 68, 96, 105, 112
ethos, 189
Europe, 17, 21, 27, 31, 41, 45, 48, 52, 56, 63, 68, 70, 74, 76, 78, 85, 90, 93, 98, 100, 104, 110, 116, 119, 121, 124, 132, 134, 140, 152, 156, 159, 161, 164, 169, 176, 178, 185, 187, 190, 194
Evolution, 1, 3, 5, xi, 21, 39, 42, 46, 51, 57, 79, 97, 152, 157, 170, 177, 180, 183, 188, 190, 192, 196
exchange, 12, 26, 38, 42, 49, 56, 75, 84, 95, 101, 104, 112, 114, 117, 119, 121, 123, 125, 129, 132, 134, 136, 148, 159, 162, 165, 169, 171, 181, 185, 189, 197
  kula exchange, 104
experiment, 2, 4, 6, 12, 30, 52, 94, 111, 113, 118, 152, 171
farmers, 27, 50, 63, 71, 74, 79, 82, 84, 87, 97, 107, 193
farming, 25, 37, 50, 52, 56, 59, 62, 70, 75, 78, 81, 85, 88, 101, 106, 108, 110, 114, 125, 148, 154, 157, 176, 180, 184, 188, 193
farming societies, 25, 62, 78, 82, 85, 101, 106, 108, 115, 125, 148, 188
fighting, 189
fistful, 152, 155
Frege, 1
geometric, x, 14, 17, 35, 67, 76, 83, 97, 149
Geröllfingen, xiii, 141, 146
gifts, 112, 194
Gilgamesh, 156
Göbekli Tepe, 97, 174
gold, 106, 121, 185
grammar, xiii, 128, 173, 183, 196
graves, 18, 41, 62, 81, 90, 101, 114, 118, 120, 141, 149, 155, 158, 180, 185, 188, 191
Greece, x, 52, 62, 64, 121, 183
Habermas, 6, 129, 169, 185
Heidegger, 28, 171
hoards, 130, 190, 193
Hof, 76
Homer, 190
Hopi, 11, 67
house communities, 65
household, 37, 41, 55, 62, 64, 66, 75, 83, 87, 96, 98, 110, 126, 157, 176
  household economy, 37

human body, 1, 4, 15, 18, 22, 27, 33, 49, 59, 69, 82, 88, 90, 92, 95, 98, 102, 126, 142, 144, 156, 170, 183, 186, 196
  bodies as calculators, 1, 15
  embodiment, xii, 28, 165, 179, 183
  hunting, 15, 17, 19, 37, 54, 56, 68, 73, 78, 81, 102, 104, 108, 156, 175, 180, 184
ideology, 65, 74, 85, 96, 104, 189, 191
idiosyncratic, 128, 159, 184
indigenous tradition, 74
individualization, 164, 195
inequality, 158, 169, 190
Kant, 6
Karanovo, 62, 64, 77, 88
Kendall, 86, 138, 150, 152
Kripke, 5, 95, 187
  stream of life, 5
Krzemionki Opatowskie, 116
Lacan, 168
  excess, 132, 169, 188
  reference frame, 167
  phantasy frame, 168
Lakoff, 23, 27, 172
language, 4, 22, 32, 35, 42, 48, 58, 60, 75, 101, 128, 168, 170, 173, 182, 184
  language determinism, 11, 14
Lèvy–Bruhl, 3
lifestyle, 70, 82, 110, 183, 189
linearity, xii, 77, 80, 84, 97, 178, 182, 197
  linear, xii, 31, 58, 71, 76, 80, 84, 88, 90, 97, 106, 109, 174, 176, 178, 182, 197
  linear measure, xii, 31, 90, 179
  lines, xii, 17, 35, 59, 64, 67, 70, 76, 80, 84, 90, 92, 110, 114, 147, 162, 176, 178, 182
  monumental line, 79, 197
long house, 71, 74, 77, 82, 84, 102, 178, 193, 195
luxury goods, 38, 40, 181

macrocosm, 69
macrolithic, xiii, 27, 39, 56, 99, 121, 128, 134, 140, 145, 159, 163, 165, 169, 172, 180, 182, 185, 193, 195
markings on bones, 18
material culture, xi, 16, 30, 32, 35, 59, 95, 131, 150, 155, 167, 174, 177, 184, 188, 190, 192
mathematics, 1, 3, 8, 10, 15, 22, 24, 28, 32, 44, 48, 51, 58, 165, 167, 173, 183
  abstract reasoning, 8
  algebra, 3
  arithmetic, x, 16, 20, 27, 32, 44, 59, 95, 98, 169
  arithmetic operations, xi, 24, 32, 170
  logical operations, 1
  logical thinking, 2
  mathematic relationships, 96
  mathematical, 1, 4, 10, 20, 30, 32, 42, 46, 48, 51, 59, 86, 98, 100, 119, 121, 135, 140, 145, 148, 153, 156, 165, 167, 178, 187, 195
  mathematical abilities, 4, 6, 16, 21, 32, 34, 60
  mathematical concepts, 10, 15, 17, 20, 27, 48, 59, 98, 100, 119, 145, 153, 156, 169, 173, 178, 188, 196
  mathematical laws, x, 33
  mathematical perception, xii, 23, 48, 156, 169, 187, 195
  mathematical rules, x, 16, 25, 43, 59, 100, 174
  mathematical thinking, x, 24, 28, 32, 42, 51, 98, 121, 168
  mathematization, 3, vi, xi, xiv, 62, 98, 100, 192, 198
  operational stage, 3
  preoperational stage, 2
  subitizing, 12, 22, 188
McLuchan, 51
  the medium is the message, 127

measure, 7, 11, 20, 27, 31, 40, 43, 50, 80, 86, 93, 110, 118, 128, 130, 132, 134, 138, 145, 148, 150, 152, 155, 159, 164, 169, 179, 183, 192, 195, 197
    lengyel fathom, 90
    measurable, 8, 159, 185, 190
    measurable accumulation, 131
    measurement, 29, 44, 86, 94, 97, 118, 134, 138, 152, 160, 165, 169, 171, 187, 198
    measurement system, 45, 88, 135, 169
    measuring, 9, 19, 27, 37, 44, 47, 63, 85, 94, 100, 108, 119, 126, 128, 134, 139, 145, 148, 152, 164, 169, 183, 186, 192, 195
    metrological, 8, 12, 22, 44, 50, 86, 95, 110, 131, 137, 146, 148, 150, 153, 162, 168, 173, 187, 192, 196
    metrological structures, xiii, 86, 138, 153, 170
    Neolithische Länge, 87
    pace measure, 87
    proportion, 12, 53, 83, 86, 124, 128, 135, 137, 141, 145, 153, 169, 171
    weighing, 143, 158, 188, 192
    weight, xiii, 28, 41, 46, 92, 94, 100, 107, 131, 135, 140, 142, 148, 159, 162, 172, 186, 190, 197
megaliths, 78, 81, 84, 92, 117, 125, 193, 195
    Calden, 83
    circular structures, 78
    cromlechs, 78
    dolmens, 78, 114
    enclosures, 77, 82, 84, 90, 120, 147, 176, 178, 193, 195
    Kuyavian barrows, 79
    Langbetten, 79
    megalithic, xii, 27, 77, 82, 85, 94, 98, 100, 114, 117, 125, 150, 154, 174, 178, 181, 190, 193
    megalithic yard, xiii, 86, 94
    megalithic structures, xii, 174, 182
    megalithism, 78, 81
    menhirs, 78, 115
    monumental constructions, 64, 73, 82, 87, 195
    monumentality, 73, 80, 98, 102, 193
    mould–stone mounds, 78
    Newgrange, 92, 179
    passage graves, 81, 114
    Passy, 85
    Stonehenge, 81, 92, 176, 179
    tombs, 53, 78, 85, 94, 101, 114, 117, 119, 125, 178, 189, 193
Merleau–Ponty, 28
Mesolithic, 73, 78, 81, 102, 104
metal, xiii, 28, 38, 40, 94, 99, 106, 119, 121, 127, 129, 136, 140, 145, 157, 171, 185, 197
metallurgy, xiii, 28, 38, 99, 106, 122, 129, 131, 134, 140, 157, 159, 161, 172, 185, 188
metaphors, xii, 22, 24, 27, 29, 34, 42, 99, 115, 159, 168, 172, 177, 183, 190
    object collection metaphor, xii, xiv, 25, 60, 169
    digit metaphor, 33, 42, 45, 60, 170, 172
    measuring stick metaphor, xii, 27, 87, 91, 95, 100, 108, 119, 126, 128, 134, 145, 148, 153, 156, 164, 169, 172, 174, 183, 186, 188, 192, 195, 198
    metaphoric blend, 24
metrologization, 168, 188, 198
microcosm, 69, 75
Middle Ages, 49, 182
miniature models, 67, 89
mining, 17, 32, 42, 87, 110, 116, 138

money, 8, 137, 185, 192
Mr. Blade, 169
myths, 26, 68, 72, 77, 81, 85, 96, 98, 100, 105, 114, 127, 156, 167, 171, 175, 180, 182, 184
narration, 10, 13, 16, 26, 48, 56, 76, 81, 96, 100, 104, 108, 119, 125, 127, 148, 169, 171, 183, 187, 197
narrative value, 104, 108, 119, 131, 187
Near East, xi, xiv, 21, 27, 35, 47, 50, 56, 59, 62, 74, 78, 88, 97, 100, 147, 156, 169, 173
Neolithic, xii, 20, 26, 33, 35, 40, 50, 52, 54, 60, 62, 67, 70, 76, 81, 84, 86, 108, 114, 119, 123, 130, 137, 140, 145, 148, 152, 167, 182, 187, 190, 193
    houses, 27, 41, 52, 62, 70, 76, 85, 96, 101, 112, 126, 175, 177, 181, 193, 195
number,
    abstract number, xi, 100, 145, 165
    concept of number, 4, 9, 16, 20, 22, 27, 33, 35, 43, 45, 50, 59, 100, 128, 134, 172, 188, 195
    number–shapes, 183
    numerals, 10, 12, 32, 42, 48
    numerical, 4, 8, 11, 22, 28, 31, 44, 51, 55, 59, 88, 91, 110, 119, 131, 139, 145, 148, 154, 156, 169, 173, 187, 195
    numerical grammar, 9
    numerosity, xiv
    perception of numbers, xii
    protonumber, 100
numbers, 3, 8, 24, 27, 29, 31, 35, 43, 47, 51, 58, 98, 100, 106, 108, 134, 141, 145, 155, 170, 185, 188
objectification, 167
ontogenesis, 1, 3
Oppenheim, 35
ornaments, 105, 121, 136, 186, 189
Palaeolithic, 10, 15, 17, 25, 57, 63, 81, 96, 128, 131, 157, 167, 171, 180, 184, 187, 196
paradox, 7, 9, 98, 167
pastoral, 70, 156, 194
Peano, 1
philosophy, x, xii, 6, 27, 167
phylogenesis, 1
Piaget, 1, 16, 25
Pirahã, 10, 14, 50
Plato, x, 30, 33, 182
Polgár, 104, 122, 126, 146
portion, vi, ix, 12, 40, 53, 83, 86, 124, 128, 134, 137, 141, 145, 150, 152, 162, 169, 171, 187, 197
prehistoric, 3, 16, 21, 27, 33, 46, 49, 51, 90, 92, 110, 116, 121, 152, 172, 197
prehistory, 1, 3, 7, 11, 17, 22, 27, 35, 51, 69, 79, 121, 156, 166, 183, 185
prestige, 74, 106, 108, 112, 122, 129, 134, 160, 180, 186, 189
primates, 7
psychology, xii, 6
    behaviour, x, 2, 6, 74, 93, 124, 149, 154, 157, 160, 175, 188, 195
    clinical experiments, 2
    egocentric thinking, 2
    egocentrism, 2
    embodied mind, 28, 30
    individuality, 51, 107, 194
    psychological interview, 2
    sociocentrism, 2
    spontaneous convictions, 2
pueblo, 64
Putnam,
    Twin Earth Model, 5
Pythagoras, x, 4, 86, 183
rank, 45, 101, 105, 169, 185, 194
rational communication methods, 169
rationality, 8

rationalization, xiii, 51, 55, 129, 180, 183
rationalized value, 132
recording, 14, 17, 33, 36, 39, 42, 45, 51, 57, 84, 86, 98, 152, 169, 178, 181, 190
religion, 69, 78, 96, 194
ritual, 8, 15, 18, 27, 34, 42, 51, 55, 75, 78, 81, 90, 96, 98, 100, 102, 112, 119, 121, 125, 147, 149, 156, 165, 175, 178, 180, 182, 188, 193
Rosetta stone, 145
Russel, 1
sacrum, 51, 93, 175
Schmandt–Besserat, 21, 36, 39, 46, 50, 53, 56, 59, 181
sedentary, 63, 70, 80, 149, 183
Seeberg, 135, 139, 146
settled life, 74
settlement agglomeration, 176
settlement organization, 64
settlement structure, 67, 97
settlements, xii, 35, 52, 56, 62, 79, 82, 88, 93, 97, 101, 109, 112, 116, 125, 129, 135, 149, 155, 168, 175, 181, 193, 195
shaman, 15, 18, 26, 175, 184
social control, 63, 75, 81, 97, 175, 178, 181
social differences, 108, 126, 158, 167
social relations, 21, 24, 26, 34, 48, 51, 57, 59, 74, 97, 113, 119, 128, 131, 141, 155, 164, 168, 171, 187
social status, 119, 192
social structure, 68, 74, 132, 158, 167, 169, 182, 191, 193, 197
socialization, 6
specialists, 43, 56, 69, 93, 96, 112, 178, 181, 192
spondylus, 104, 108, 181
staple finance, xiv
Stone Age, 133, 145, 189
storytelling, 10, 128

symbolism, 3, 9, 16, 20, 28, 34, 40, 51, 55, 63, 67, 72, 77, 84, 87, 96, 99, 102, 104, 110, 114, 120, 122, 127, 131, 154, 156, 165, 169, 172, 174, 177, 182, 184, 190, 195
symbols, 3, 9, 16, 20, 24, 28, 34, 40, 43, 51, 55, 81, 99, 108, 110, 154, 174, 179, 195
symmetry, 69
symposia, 156
Technology, xi, 24, 28, 62, 99, 106, 119, 129, 131, 140, 145, 171, 186, 193
tectomorphs, 65, 67, 72
tells, 38, 65, 77, 101
theatres of memory, 168
Thom, 86, 89, 94, 138, 150, 172, 179, 194
Tiszapolgár–Basatanya, 122, 127
tokens, xii, 21, 33, 47, 50, 56, 62, 100, 170, 172, 174, 178, 180
Tomasello, 6
  scenes of joint attention, 7
  verb islands, 7
tools, 6, 17, 27, 39, 68, 72, 83, 94, 98, 101, 106, 111, 114, 117, 121, 124, 127, 131, 148, 157, 159, 165, 169, 172, 180, 183, 189, 193
  axes, xiii, 40, 94, 102, 104, 106, 129, 132, 150, 158, 185, 189, 191, 193, 195
  blades, xiii, 56, 104, 109, 121, 127, 140, 163, 165, 169, 186, 193, 195
  flint, xiii, 18, 41, 56, 104, 106, 108, 114, 122, 125, 127, 129, 132, 140, 145, 158, 160, 164, 171, 182, 193, 196
  flint axe, 109, 125
  flint blade, xiii, 109, 122, 196
  flint nodules, 123
  fragmentation, of, xiii, 57, 97, 105, 124, 126, 128, 131,

159, 163, 169, 172, 177, 181, 187, 196
long blades, 121
macrolithic blades, 56, 121, 128, 140, 163, 165, 169, 186, 195
macrolithic industries, xiii, 39, 169, 174, 195
macrolithic technology, xiii, 129, 172
macrolithization, xiii, 109
macroliths, xiii, 126, 131, 159, 169, 172, 180, 183, 187, 196
stone axe, 106, 108
tool production, xiii, 196
tuff axes, 120
topos, 26
transcendental, x, 30, 68
trisected building, 177
Ubaid period, 42, 178
urbanization, 38, 42, 178
Uruk period, 47

Varna, 101, 105, 122, 185
Vocabulary, 9, 11, 33, 51, 115, 128, 132, 134, 140, 145, 165, 169, 172, 182, 196
warrior, 93, 102, 156, 177, 189
wealth, xiv, 39, 106, 122, 131, 178, 181, 183, 186, 189, 196
wealth finance, xiv, 181
Whorf, 10, 32, 67
wildness, 75, 177
Wittgenstein, 4, 95
   conditions of affirmability, 5
   motley collection of language games, 4
   Philosophical Investigations, 5
   Tractatus Logico-Philosophicus, 5
women, 75, 103, 106, 118, 122, 152, 155, 177
Zeilensiedlung, 76